Asphalt Cold Mix Manual

Asphalt Institute
Manual Series No. 14 (MS-14)
Third Edition

**U.S. Library of Congress Catalog Card No. 89-62536
Copyright © 1989 The Asphalt Institute**

All reasonable care has been taken in the preparation of this Manual; however, the Asphalt Institute can accept no responsibility for the consequences of any inaccuracy which it might contain. The principles and guidelines presented herein are to be used and interpreted by qualified engineers.

Photographs and drawings of equipment used in this publication are for illustration only and do not imply preferential endorsement of any particular manufacturer by the Asphalt Institute.

This publication incorporates dual units of measurement: the International System of Units, SI (metric), followed by U.S. Customary Units.

PRINTED IN USA

FOREWORD

This publication has been prepared for the purpose of assisting and guiding engineers in the analysis, design, and control of asphalt cold mix construction.

The text brings together in a concise and logical relationship typical mixes that are recognized and proven satisfactory for preparation by cold mix methods. Also, with this information engineers will be able to adapt specific conditions to the general specifications shown herein. Assistance with the application of these specifications and construction procedures is readily available from the Asphalt Institute engineering staff.

Contents

	Page
List of Asphalt Institute Member Companies	ii
Foreword	v
List of Tables	ix
List of Illustrations	xi

Part I: Basic Information

CHAPTER I: INTRODUCTION	1
CHAPTER II: MATERIALS	3
A. Asphalt	3
B. Aggregates	6
CHAPTER III: MIX DESIGN	9

Part II: In-Place Mixing Equipment

CHAPTER IV: EQUIPMENT FOR MIXED-IN-PLACE	11
A. Mixing Equipment	11
B. Spreading Equipment	14
C. Compacting Equipment	16
CHAPTER V: CONSTRUCTION	20

Part III: Central Plant Mixing

CHAPTER VI: EQUIPMENT FOR PLANT MIXES	29
CHAPTER VII: CENTRAL PLANT MIX CONSTRUCTION	32

Part IV: Appendices

APPENDIX A: SUGGESTED PLANT MIX GUIDELINES	35
Guideline PM-1: Cold Asphalt Plant Mix	
A. General Requirements	35
B. Materials	36
C. Construction	41

	Page
APPENDIX B: SUGGESTED MIXED-IN-PLACE GUIDELINES	45
Guideline RM-1: Mixed-in-Place Courses	
A. General Requirements	45
B. Materials	48
C. Construction	48
Guideline RM-2: Road-Mixed Asphalt Courses for Base and Surface (Sand or Soil)	
A. General Requirements	52
B. Materials	54
C. Construction	55
APPENDIX C: SUGGESTED GUIDELINES FOR STOCKPILE PATCHING MIXTURES	58
Guideline PM-2: Plant-Mixed Asphalt Stockpile Maintenance Mixtures	
A. General Requirements	58
B. Materials	59
C. Mixture Preparation	60
Guideline RM-3: Mixed-in-Place Asphalt Stockpile Maintenance Mixtures	
A. General Requirements	62
B. Materials	63
C. Construction	64
APPENDIX D: RANDOM SAMPLING PLANS	66
APPENDIX E: MODIFIED HVEEM METHOD FOR EMULSIFIED ASPHALT-AGGREGATE COLD MIXTURE DESIGN	71
APPENDIX F: MARSHALL METHOD FOR EMULSIFIED ASPHALT-AGGREGATE COLD MIXTURE DESIGN	106
APPENDIX G: MODIFIED HVEEM METHOD FOR CUTBACK ASPHALT-AGGREGATE COLD MIXTURE DESIGN	123
APPENDIX H: MARSHALL METHOD FOR CUTBACK ASPHALT-AGGREGATE COLD MIXTURE DESIGN	144
APPENDIX I: MISCELLANEOUS TABLES	162
Index	177
List of Asphalt Institute Engineering Offices	187

List of Tables

Table		Page
II-1	Guide for Uses of Asphalt in Cold Mix	4
II-2	Typical Asphalt Temperatures for Cold Mix Construction	7
D-1	Random Numbers for General Sampling Procedure	69
E-1	Selection of Emulsified Asphalt Content	74
E-2	Surface Area Factors	76
E-3	Variables Affecting Asphalt Dispersion	83
E-4	Multiplying Factors for Cohesiometer Values	103
E-5	Design Criteria For Emulsified Asphalt-Aggregate Mixes	105
F-1	Emulsified Asphalt Mixture Data Sheet	116
F-2	Stability Correlation Ratios	119
F-3	Emulsified Asphalt-Aggregate Mixture Design Criteria	121
G-1	Suggested Criteria for Cutback Asphalt Mixes	143
H-1	Marshall Design Criteria For Paving Mixtures Containing Cutback Asphalt	156
H-2	Characteristics of Mineral Aggregate	158
H-3	Analysis of Medium Curing Asphalts	158
H-4	Characteristics of Paving Mixtures Prepared With MC-250	160
I-1	Typical Temperatures for Uses of Cutback and Emulsified Asphalt—Degrees Celsius	162
I-2	Typical Temperatures for Uses of Cutback and Emulsified Asphalt—Degrees Fahrenheit	163
I-3	Temperature-Volume Corrections for Asphalt Materials—Degrees Celsius	166
I-4	Temperature-Volume Corrections for Asphalt Materials—Degrees Fahrenheit	168
I-5	Temperature-Volume Corrections for Emulsified Asphalts	170
I-6	Percent Capacities for Various Depths of Cylindrical Tanks in Horizontal Positions	172

Table		Page
I-7	Weights and Volumes of Asphalt Materials (Metric Units)	173
I-8	Weights and Volumes of Asphalt Materials	173
I-9	Weight per Cubic Metre of Dry Mineral Aggregates of Different Specific Gravity and Various Void Contents	174
I-10	Weight per Cubic Foot and per Cubic Yard of Dry Mineral Aggregates of Different Specific Gravity and Various Void Contents	175

List of Illustrations

Figure		Page
IV-1	Rotary mixer with asphalt supply tank	12
IV-2	Rotary mixer	12
IV-3	Planer/rotary mixer operation	13
IV-4	Blade mixing	15
IV-5	Hopper-type pugmill travel plant operating from a windrow	15
IV-6	Asphalt distributor	16
IV-7	Paver operating from windrow with a pick-up loader	17
IV-8	Pneumatic-tired roller	18
IV-9	Steel-wheeled tandem roller	18
IV-10	Vibratory roller	19
V-1	Measurements for determining windrow quantities	22
V-2	Spreading and compacting train	27
VI-1	Stationary cold-mix plant (batch)	30
VI-2	Flow diagram of a typical cold mix continuous plant	30
VI-3	Cold mix continuous plant	31
VII-1	Cold plant mix being loaded into haul-truck	33
D-1	Schematic diagram illustrating lot, sample, subsample, and sample unit	67
E-1	Testing schedule for dense-graded emulsified asphalt mixes	72
E-2	Chart for determining surface constant for fine material, K_f, from C.K.E., Hveem method of design	79
E-3	Chart for determining surface constant for coarse material, K_c, from coarse aggregate absorption, Hveem method of design	80
E-4	Chart for combining K_f and K_c to determine surface constant for combined aggregate, K_m, Hveem method of design	81
E-5	Chart for computing oil ratio for dense-graded asphalt mixtures, Hveem method of design	82

Figure		Page
E-6	Apparatus for Hveem C.K.E. tests	84
E-7	Transfer of mix to mold	88
E-8	Rodding mix in mold	88
E-9	Mechanical kneading compactor	88
E-10	Resilient modulus device	91
E-11	Transducers and M_r yoke	91
E-12	M_r yoke on holder	92
E-13	Tightening M_r clamping screws	92
E-14	Seating M_r specimen on loading block	92
E-15	Adjusting M_r recording meter	93
E-16	Adjusting M_r pressure regulator	93
E-17	Vacuum manometer and desicator	95
E-18	Hveem stabilometer	97
E-19	Chart for determining R-value from stabilometer data	99
E-20	Chart for correcting R-value	100
E-21	Chart for correcting stabilometer values	102
E-22	Diagram showing principal features of the Hveem cohesiometer	104
F-1	Typical emulsified asphalt-aggregate mixture design plots	122
G-1	Chart for correcting asphalt requirement due to increasing viscosity of asphalt	126
G-2	Transfer of mix to mold	131
G-3	Rodding mix in mold	131
G-4	Swell test apparatus	134
G-5	Diagram showing principal features of Hveem stabilometer	135
G-6	Hveem stabilometer	135
G-7	Diagram showing features of Hveem cohesiometer	136
G-8	Aluminum seal cap	136
G-9	Pressing standard	137
G-10	Chart for correcting stabilometer values	140

Figure		Page
G-11	Moisture vapor susceptibility test	141
H-1	Example of viscosity-temperature relationship for medium curing cutback asphalts made from same base stocks	148
H-2	Example of composition chart: medium-curing cutback asphalt made from the same base stocks	149
H-3	Example of specific gravity at 25° C (77° F) of medium-curing cutback asphalts made from the same base stocks	152
H-4	Test property curves for cutback asphalt mixture design data by the Marshall Method	154
H-5	Minimum percent voids in mineral aggregate (VMA)	157
H-6	Example of aggregate gradation showing deviation from maximum density curve	159

Part I: Basic Information

Chapter I. Introduction

1.01 ASPHALT COLD MIX.—Asphalt cold mix is a mixture of unheated mineral aggregate and emulsified or cutback asphalt. Classified by the method of mixing, there are two types of cold mix: plant-mix and mixed-in-place (road mix).

Plant-mixed cold mixes are produced in stationary plants that permit close control of the production process from materials proportioning through mixing. Spreading and compacting is done with conventional paving equipment.

Mixed-in-place cold mixes are produced at the paving site by means of travel plants, motor graders, or special in-place mixing equipment.

1.02 ADVANTAGES OF ASPHALT COLD MIXES.—
- *Versatile.* A number of types and grades of emulsified and cutback asphalt are available to satisfy the varying requirements of different aggregate and weather conditions.
- *Economical.* High production rates are possible with a comparatively low investment in equipment. Additionally, locally available aggregate can be used. As relatively little equipment is required for cold-mix construction, it is simple and economical to use cold mixes.
- *Non-polluting.* As dryers are not needed to heat the aggregate, no smoke is produced and dust emission is quite low. When emulsified asphalt is used, there are usually no objectionable fumes or odors.

1.03 LIMITATIONS OF ASPHALT COLD MIXES.—
- *Weather.* Cold mix construction should not be done when ambient temperatures under 10°C (50°F) are expected, or when generally poor weather is predicted. As the aggregate is not heated, its maximum temperature is limited to that of the atmosphere, plus that attributable to solar radiation. Upon application, the asphalt quickly reaches the temperature of the aggregate. If it is too cool, mixing is difficult. Also, extra manipulation may be required to remove volatiles in cool and humid conditions.
- *Surface Moisture.* The determination of surface moisture is based upon the surface dry weight of the aggregate as it will be used. Up to 3 (sometimes more) percent surface moisture may be required on the aggregate for successful mixing with emulsified asphalt and subsequent compacting of the mixture. If cutback asphalt is used, the surface moisture content should usually be less than 3 percent. With either type of asphalt, excessive surface moisture causes problems in mixing, compaction and curing. Only the slow-setting (SS-1 and SS-1h) and anionic grades of medium-setting (MS-1, MS-2,

and MS-2h) emulsions require moisture for mixing. The HFMS grades (particularly the HFMS-2s) and the CMS-2 and -2h emulsions, along with other available modifications, contain a quantity of petroleum distillates. These products perform much better with dry aggregates (mixing, laying, etc.), than with wet aggregate.
- *Application.* Asphalt cold mixtures may be used for surface, base, or subbase courses if the pavement structure is designed properly. As a surface course, cold mix is typically suitable for medium and light traffic but open-graded emulsion mixes have been used for heavy traffic. For base or subbase, it is suitable for all types of traffic.
- *Quality Control.* Excellent pavements can be achieved with mixed-in-place cold mixes when proper attention is devoted to ensuring uniformity in the quantities of aggregates, aggregate gradation, and applied quantities of asphalt. However, the production process for these mixes is generally more difficult to control than for plant-mixed cold mixes.

1.04 PREPARATION OF ROADBED.—Every part of the structure contributes to the effective performance of the completed pavement. Careful shaping and compacting of the roadbed is the first step in building a sound pavement structure.

An important part of this process involves detecting and repairing or replacing unstable areas with sound material equal in quality to the rest of the roadbed.

For overlays, holes and depressions in existing granular and asphalt surfaces should be repaired by patching; bumps, waves, and corrugations should be removed; and excess asphalt in cracks and patches should be removed.

A properly designed and constructed surface drainage system is necessary to minimize infiltration of water into the pavement structure. Occasionally, subsurface drains may be needed at some locations to ensure continuous removal of water from the structural elements of the pavement.

The final steps in roadbed preparation are compacting the foundation to the specified density, whether new location or existing aggregate surface, and shaping it to the required line, grade, and cross-section. Priming the prepared roadbed with asphalt is advisable if the roadbed will be used by traffic or will be subjected to wet weather.

Chapter II. **Materials**

A. Asphalt

2.01 TYPES OF ASPHALT.—Two types of asphalts are available for cold mix operations: emulsified asphalt and cutback asphalt.

Emulsified asphalt is a dispersion of asphalt cement in water that contains a small amount of an emulsifying agent, a heterogeneous system containing two normally immiscible phases—asphalt and water. Water forms the continuous phase of the emulsion, and minute globules of asphalt form the discontinuous phase. Emulsified asphalts may be either of two types: anionic (electro-negatively charged asphalt globules) or cationic (electro-positively charged globules), depending upon the emulsifying agent. These two types are further classified by the rate of setting, or breaking, of the emulsion.

There are thirteen grades of medium and slow setting (MS and SS) emulsified asphalts that are used for mixing with aggregates. The anionic grades are MS-1, MS-2, MS-2h, HFMS-1, HFMS-2, HFMS-2h, HFMS-2s, SS-1 and SS-1h. The cationic grades are CMS-2, CMS-2h, CSS-1, and CSS-1h. Some user agencies also specify CMS-2s grade. The numeral in the designation is an indication of the relative viscosity of the emulsion. For example, the MS-2 grade is a more viscous material than MS-1. The actual viscosity ranges are given in the specifications. Standard specifications for cationic emulsified asphalts are found under American Society of Testing Materials (ASTM) Designation D 2397 or American Association of State and Highway Transportation Officials (AASHTO) Designation M 208 and for other emulsified asphalts under ASTM Designation D 977 or AASHTO Designation M 140.

Cutback asphalt is asphalt cement that has been liquefied by blending with petroleum solvents (also called diluents). Upon exposure to atmospheric conditions the diluents evaporate*, leaving the asphalt cement to perform its function.

Cutback asphalts are graded by the type and amount of solvent used to make them liquid, thereby enabling them to be handled at lower temperatures. For example, medium curing (MC) cutback asphalts (ASTM D 2027 or AASHTO M 82) are produced by blending asphalt cement with a kerosene-type solvent. Slow curing (SC) cutback asphalts (ASTM D 2026) are made by reducing the crude oil directly to grade by distillation, or by fluxing a paving asphalt with a light oil.

Cutback asphalts also are given numerical suffixes to denote viscosity; specifically, the minimum specified kinematic viscosity, measured in centistokes at 60°C (140°F). For example, ASTM specifies that the minimum kinematic viscosity of MC-250 at 60°C (140°F) will be 250 centistokes.

2.02 SELECTING THE PROPER ASPHALT.—Selection of the proper type and grade of asphalt material to use for each project is most important. First

---

*Because of diluent evaporation, environmental regulations may restrict or prohibit the use of cutbacks in many areas.

consideration should be given to the type and grade performing most satisfactorily on local projects with aggregate gradations and traffic conditions similar to those on the project under study. Additionally, the standard asphalt specifications serve to assist in selecting the proper material for each project by providing a clear, meaningful comparison between types and grades. Beyond the specifications, however, independent, personal judgment must still be exercised in making the selection. The decision must consider usage of the completed pavement, environmental conditions at the pavement location, type of equipment available and construction operations. This judgment is founded upon three broad engineering considerations: properties of the residual asphalt; consistency; and curing or setting rate.

While general guidelines can be given for selecting emulsions or cutbacks, laboratory testing is strongly recommended. There is no good substitute for a laboratory evaluation of the asphalt and the aggregate to be used. Different types and quantities of asphalt should be tested with the aggregate to find the best combination for the intended use. An experienced technician can determine the type and amount of cutback to be used. The technician can also determine if, in the case of emulsions, additional water must be added, and the amount of time for breaking to occur.

Table II-1 shows the various types and grades of asphalt materials typically used for cold-mix construction. Certain state or locally specified grades or designations may be used where appropriate.

Table II-1. Guide for Uses of Asphalt in Cold Mix

Type of Construction	Emulsified Asphalts [1]									Cutback Asphalts						
	Anionic					Cationic				Medium Curing (MC) [2]				Slow Curing (SC)		
	MS-2, HFMS-2	MS-2h, HFMS-2h	HFMS-2s	SS-1	SS-1h	CMS-2	CMS-2h	CSS-1	CSS-1h	70	250	800	3000	250	800	3000
Cold-Laid Plant Mix [3] Pavement Base and Surfaces																
Open-Graded Aggregate	X	X				X	X									
Well-Graded Aggregate		X	X	X				X	X		X	X	X	X	X	X
Patching, Immediate Use			X	X				X	X		X	X			X	
Patching, Stockpile											X	X		X	X	
Mixed-in-Place (Road Mix) [3] Pavement Base and Surfaces																
Open-Graded Aggregate	X	X				X	X					X	X		X	X
Well-Graded Aggregate		X	X	X				X	X		X	X		X	X	
Sand		X	X	X				X	X	X	X	X				
Sandy Soil		X	X	X				X	X		X	X				
Patching, Immediate Use		X	X	X				X	X		X	X			X	
Cold-Mix Recycling [3]	X	X	X	X	X	X	X	X	X							

[1] Emulsified asphalts shown are AASHTO and ASTM grades and may not include all grades produced in all geographical areas.
[2] Before using MC's for spray applications check with local pollution control agency.
[3] Evaluation of emulsified asphalt-aggregate system required to determine the proper grade of asphalt to use.

Properties of Residual Asphalt. Generally desired is a residual asphalt that will provide the greatest cohesion in the final mix. But other considerations are the type of aggregate available and the firmness of the subgrade.

A long standing field rule states that one should use the heaviest (most viscous) asphalt that can be readily worked. Applying that rule means that the type of mixing equipment to be used is a big factor in establishing the required grade of asphalt.

Consistency. Viscosity, a measure of consistency, is the property of a fluid that resists the force tending to cause the fluid to flow. The viscosity of asphalt varies with temperature, becoming more viscous as temperature decreases. As the asphalt used in cold mixes must be readily workable at temperatures encountered during construction, the viscosity of various asphalt materials at ambient temperatures should be considered. Thus, ambient temperature is a key factor in selecting a viscosity of asphalt that will produce desirable, uniform mixes.

Another factor governing the consistency of the asphalt selected for any given project is the gradation of the aggregates. As a generalization, open-graded mixes require a more viscous binder than that required for well-graded mixes. In either case, it is desirable to use the most viscous grade that can be handled during construction. When a mix has a high proportion of fine material passing the 75 μm (No. 200) sieve, mixing is usually more difficult, and medium to low viscosity asphalts may be more effective. With fewer fines passing the 75 μm (No. 200) sieve, mixing is somewhat easier and a higher viscosity asphalt may be used. Furthermore, in open-graded mixes, the use of higher viscosity asphalts will reduce the possibility of asphalt drain off from the aggregate particles.

Rate of Curing—Emulsified Asphalt. Emulsified asphalts depend on the evaporation of water for development of their curing and adhesion characteristics. Water displacement can be fairly rapid under favorable weather conditions but high humidity, low temperatures, or rainfall soon after application can deter proper curing. Although atmospheric conditions are less critical for cationic emulsions than for anionic, they still depend on weather conditions for optimum results.

Consideration of climatic conditions anticipated during construction is important when using emulsified asphalts. The general rule is to select equipment which provides for rapid construction. Central mix plants are preferred, followed closely by travel plants. High production rates can be achieved in good weather and the project can easily be stopped in inclement weather. Mixed-in-place construction (such as blade mixing) is least desirable since large areas are exposed during mixing. Rain could increase the moisture content above optimum, requiring aeration. On the other hand, the suitability of the central plant hinges on availability of the plant within an economic haul distance.

When the MS and SS grades are used for cold mixes, the use of slightly damp aggregates facilitates the mixing and coating process. The development of strength in the SS types depends mainly on dehydration and absorption, with removal of water by either of these mechanisms breaking the emulsion. Solvent-free CMS and CSS emulsions require that the moisture on the aggregate be at or near optimum for proper mixing and coating.

Some types of emulsions contain slight amounts of petroleum solvent to aid in the mixing and coating process. While the solvent does not enter directly into the breaking mechanism, provisions must be made for the evaporation of the solvent in order for the mixture to be properly cured. Where multiple courses are to be placed, a successive course should not be applied until the water (and solvent, if applicable) has been removed from the preceding course.

Medium-Setting Emulsions—The medium-setting grades are designed for mixing with coarse aggregate. Because these grades do not break immediately upon contact with the aggregate, mixes using them remain workable for a short time.

High-float medium-setting asphalt emulsions may give better aggregate coating and asphalt retention under extreme temperature conditions. They may be used with coarse- or dense-graded aggregates.

Slow-Setting Emulsions—The slow-setting grades are designed for maximum mixing stability. They are used with high fines content, dense-graded aggregates. All slow-setting grades have low viscosities that can be further reduced by adding water.

Rate of Curing—Cutback Asphalt. The quantity of asphalt applied, type and grade of asphalt, humidity and wind, the amount of rain, and the range of ambient temperatures of the region during application affect the relative curing rates of cutback asphalts. Both the rate of curing and the hardening of the asphalt (after the petroleum diluent has evaporated) are affected not only by the local climate but also by the mixing temperature. An inter-relationship of all these factors exists and must be taken into account in selecting a specific asphalt for a specific purpose.

Generally for cutback asphalt, the lighter the solvent, the faster it will evaporate; but if more of it is present, the longer it will take to cure. Further, the colder the atmospheric temperature and the higher the humidity, the slower will be the rate of cure.

2.03 GUIDE TO USES OF ASPHALT.—Table II-1 is a guide for selecting types and grades of asphalt materials for cold mixes. Table II-2 can be used to select a range of temperatures for the proper spraying and mixing of emulsified and cutback asphalts.

B. Aggregates

2.04 AGGREGATES.—A wide variety of aggregates and soil-aggregate combinations, ranging from well-graded crushed rock to silty sands, can be cold mixed satisfactorily with asphalt emulsions or cutbacks. Factors such as shape of aggregate particles, type and amount of fines, and differences in specific gravities of the mineral aggregates must be taken into account in establishing the controls needed to ensure successful results.

Commercially crushed rock, slag, and gravel are widely available for cold mixes. There are also many localities where it is economical to crush aggregate

Table II-2. Typical Asphalt Temperatures for Cold Mix Construction

Type and Grade	Asphalt Temperature When the Asphalt is Metered through a Mixing Facility	Asphalt Temperature When the Asphalt is Applied to a Windrow Prior to Mixing
Emulsified Asphalts Anionic MS-1, MS-2, MS-2h SS-1, SS-1h HFMS-2, HFMS-2h, HFMS-2s Cationic CMS-2, CMS-2h CSS-1, CSS-1h	10-70°C (50-160°F)[1] 10-70°C (50-160°F)[1]	20-70°C (70-160°F) 20-70°C (70-160°F)
Cutback Asphalts[2] MC, SC Asphalts 70 250 800 3000	 — 55-80°C (135-175°F)[3] 75-100°C (165-210°F)[3] 80-115°C (175-240°F)[3]	 20°C+ (65°F+)[4] 40°C+ (105°F+)[4] 55°C+ (135°F+)[4] —

NOTES:
[1] Temperature of the emulsified asphalt in mixing facility.
[2] Application temperatures may, in some cases, be above the flash point of the material. Caution must therefore be exercised to prevent fire or an explosion.
[3] Temperature of mixture immediately after mixing rather than temperature of asphalt.
[4] The maximum temperature shall be below that at which fogging occurs.

from road cuts, pits, or nearby quarry sites. Crusher run material containing 0 to 10 percent passing a 75 μm (No. 200) sieve, and a maximum size of 50 mm (2 in.), (or two-thirds of the course thickness, whichever is smaller,) usually proves excellent for base construction. Generally, the entire output of the crusher that is below the specified maximum size may be used in the asphalt pavement structure; this affords the greatest economy.

In areas where aggregates meeting standard tests are scarce, it often is possible to use lesser quality materials if experience has shown them to be satisfactory or where research and testing warrants such use. Local materials ranging all the way from fine granular soils through clean sands and gravels have been used. Other materials, proved by experience, can also be used. With finer gradations, however, mixing, aeration and compaction problems may be encountered. A criterion that can be used to determine the usefulness of soil material for cold mixes is given in Article 2.05.

On many projects, by careful selection, suitable materials may be obtained from excavation or borrow. On other sites, materials may be available within or adjacent to the right-of-way either as they are in pits, by blending different strata or areas of the pits, or by blending with other materials at the pit. Well-graded, processed aggregates are always desirable for any course of the asphalt pavement structure, but many poorly-graded and gap-graded aggregates have proven adequate for base course mixes when combined with the proper asphalt using good construction procedures. But, whether for base or surface, the best quality aggregate materials readily available to the job site should be used for the pavement mixture.

2.05 AGGREGATE TESTS.—One measure of the suitability of soil material for cold-mixing is that the product of the plasticity index (ASTM D 424 or AASHTO T 90) and the percent passing the 75 μm (No. 200) sieve should be less than 72.

Another measure of suitability is the sand equivalent test. The sand equivalent test (ASTM D 2419 or AASHTO T 176) is used to detect the presence of excessive quantities of clay, plastic fines and dust. In general, materials with a sand equivalent above 35 can be stabilized successfully with asphalt. The chances of success with 20-30 sand equivalent materials depend upon the ability of the asphalt to waterproof the particles. Attempts to stabilize clay-gravels with sand equivalents of less than 20 generally are not successful. Such mixes become stiff during mixing, the fines ball up, and the asphalt either strips from or fails to coat the coarse aggregate.

The Los Angeles Abrasion Test (ASTM C 131 or AASHTO T 96) is used to measure wear or abrasion resistance of mineral aggregate. Relatively high resistance to wear, as indicated by a low percent of abrasion loss, is a desirable characteristic of aggregates to be used in asphalt pavement construction. For cold mixes, the maximum abrasion loss criterion is generally set at 40 percent.

2.06 HARD-TO-COAT AGGREGATE.—Some cold mix aggregates are hard to coat with asphalt. Generally, these aggregates are hydrophilic—more attracted to water than to asphalt—and because the aggregates are not dried of all moisture they reject the asphalt. Prolonging the mixing process in an attempt to coat them usually is futile, resulting only in loss of volatiles and a poor mixture. With such aggregates, the best way to overcome this phenomenon is to change the electrical charge of the ions either on the surface of the aggregate or in the asphalt. When using anionic emulsified asphalt or cutback asphalt, coating may be improved by incorporating small amounts of hydrated lime or antistrip additive. The coating may also be improved on certain aggregates by using cationic emulsified asphalt as a binder, without the use of hydrated lime. In other cases, improved coating may result from a change from cationic emulsified asphalt to anionic. Laboratory tests are required, however, to determine the type and quantity of material to use.

Chapter III. **Mix Design**

3.01 GENERAL.—There is no universally-accepted emulsified or cutback asphalt-aggregate mix-design method; but nearly all of those in use employ some parts or modifications of the standard Marshall (ASTM D 1559 or AASHTO T 245) or Hveem (ASTM D 1560 and D 1561 or AASHTO T 246 and T 247) test methods.

Two emulsified asphalt mix-design procedures are (1) the Illinois method based on a modified Marshall mix design procedure and a moisture durability test and (2) the Asphalt Institute's method based on a modified Hveem procedure plus a resilient modulus test. These laboratory methods were further evaluated and modified by the Institute as part of the National Cooperative Highway Research Program (NCHRP) Project 9-5, "Design of Emulsified Asphalt Paving Mixtures," (NCHRP Report 259). The methods contained in Appendix E, Modified Hveem Method For Emulsified Asphalt-Aggregate Cold Mixture Design and Appendix F, Marshall Method For Emulsified Asphalt-Aggregate Cold Mixture Design are the result of the earlier development and later evaluation by the Institute.

In 1974 the Institute first published two laboratory methods to give guidance in designing mixtures using cutback asphalts. They are also modifications to the Hveem and Marshall methods of mix design. The two methods are contained in Appendix G, Modified Hveem Method for Cutback Asphalt-Aggregate Cold Mixture Design and Appendix H, Marshall Method for Cutback Asphalt-Aggregate Cold Mixture Design.

3.02 EMPIRICAL FORMULAS.—If the laboratory equipment required in the methods described in this chapter is not available, one of these formulas might be used to set the initial asphalt content, to be adjusted as needed after construction begins.

FORMULA NO. 1
Formula for Determination of Estimated Percent
Asphalt Emulsion Requirement for Dense Graded Mixes

$$P = (0.05A + 0.1B + 0.5C) \times (0.7)$$

where $P =$ Percent by weight of asphalt emulsion, based on weight of graded mineral aggregate
$A =$ Percent of mineral aggregate retained on 2.36 mm (No. 8) sieve
$B =$ Percent of mineral aggregate passing 2.36 mm (No. 8) sieve, and retained on 75 µm (No. 200) sieve
$C =$ Percent of mineral aggregate passing 75 µm (No. 200) sieve. (All percentages expressed as a whole number)

FORMULA NO. 2
Formula for Determination of Estimated Percent
MC and SC Asphalt Requirement for Dense Graded Mixes

$$P = 0.02A + 0.07B + 0.15C + 0.20D$$

where P = Percent of asphalt material by weight of dry aggregate
 A = Percent of mineral aggregate retained on 300 μm (No. 50) sieve
 B = Percent of mineral aggregate passing 300 μm (No. 50) and retained on 150 μm (No. 100) sieve
 C = Percent of mineral aggregate passing 150 μm (No. 100) and retained on 75 μm (No. 200) sieve
 D = Percent of mineral aggregate passing the 75 μm (No. 200) sieve.
 (All percentages expressed as a whole number)

Absorptive aggregates—such as slag, limerock, vesicular lava and coral—will require additional asphalt.

Example:

Given the following aggregate and construction method determine the required cutback asphalt:

Method of construction: blade mixing;

Aggregate: relatively nonabsorptive gravel, gradation shown below; and

Asphalt: MC-250

Sieve size	19.0 mm (3/4")	9.5 mm (3/8")	4.75 mm (No. 4)	2.36 mm (No. 8)	600 μm (No. 30)	300 μm (No. 50)	150 μm (No. 100)	75 μm (No. 200)
Percent passing	100	75	59	36	20	16	12	8

Solution:

$$P = 0.02 (100 - 16) + 0.07 (16 - 12) + 0.15 (12 - 8) + 0.20 (8) = 4.2$$
(use 4 percent cutback asphalt by weight of dry aggregate)

Part II: In-Place Mixing Equipment

Chapter IV. Equipment for Mixed-in-Place

4.01 GENERAL.—There are available a number of different types of machines to accomplish mixed-in-place construction. A review of the various types of mixing, spreading, and compacting equipment commonly used is presented in this chapter.

A. Mixing Equipment

4.02 ROTARY MIXERS.—Rotary or mechanical on-site mixing is accomplished by what is essentially a mobile mixing chamber mounted on a self-propelled machine. Within the chamber, usually about 2 m (7 ft) wide and open at the bottom, are one or several shafts transverse to the roadbed, on which are mounted tines or cutting blades that revolve at relatively high speed. As the machine moves ahead, it strikes off behind it a uniform course of asphalt-aggregate mixture. Most rotary mixers have a single shaft. Some single shaft mixers are equipped with a system that adds asphalt by spraying it into the mixing chamber as the machine moves ahead, with the amount of spray being synchronized with the travel speed (Figure IV-1). Other machines, however, must be used in conjunction with an asphalt distributor that sprays asphalt onto the aggregates immediately ahead of the mobile mixer (Figure IV-2).

Both types of machine have the common capability of effecting a smooth bottom cut and then blending the material and asphalt into the mixture specified. But each type is individually marked by certain devices and features that enable it to perform. Machines with built-in asphalt feeding must have the capability for accurate metering and blending of asphalt into the in-place materials in synchronization with a continuous forward movement. Furthermore, they must have spray bars that will distribute the liquid uniformly across the mixer's width. They must be equipped with controls for both depth of cutting and processing and for spreading the mixed material being laid out behind the mixing chamber.

Rotary mixers without asphalt spraying equipment generally feature controls that permit adjustment of cutting depth to at least 250 mm (10 in.), adjustment of the tail board, and adjustment of the hood itself for aeration purposes.

Some milling/planing machines can be filled with a spray/additive system. This system is calibrated for proper asphalt addition. The machine can mix the asphalt with existing road material or incorporate additional virgin aggregate that has been spread onto the road surface. The final mix can be discharged into a windrow for placement or directly into a spreader or self-propelled paver for placement. Another use of a milling/planing machine is to break-up and size an existing paved surface to be cold recycled. Then the reduced material is run through a rotary mixer where asphalt is blended and cold mix is discharged onto the road surface. (Figure IV-3).

Figure IV-1. Rotary mixer with asphalt supply tank.

Figure IV-2. Rotary Mixer.

Figure IV-3. Planer/rotary mixer operation.

4.03 MOTOR GRADERS.—Blade mixing is the on-site mixing of asphalt and in-place materials on the roadbed by a motor grader (Figure IV-4). The asphalt is applied directly ahead of the motor grader by an asphalt distributor.

For most effective blade mixing, the motor grader should have a blade at least 3m (10 ft) long, and should have a wheel-base of at least 4.5 m (15 ft). Motor graders used for final laydown and finishing of the surface should be equipped with smooth, rather than treaded, pneumatic tires. Scarifier or plow attachments may be mounted before or behind the blade, or both.

4.04 TRAVEL PLANTS.—Travel plants are self-propelled units that proportion and mix aggregates and asphalt as they move along the road. One type of travel plant receives aggregate into its hopper from haul trucks, adds and mixes asphalt, and spreads the mix to the rear as it moves along the roadbed (Figure IV-5).

Certain features and performance capabilities are common to all travel plants, enabling them to operate effectively and to produce a mix meeting design and specification criteria. The tracks or wheels on which the machine moves must be sized, designed and positioned so that they do not damage or rut the surface on which it operates when the plant is fully loaded. The basic purpose of the travel plant is to mix asphalt and aggregate. Some machines are equipped with devices that maintain the proper proportions automatically. Others, however, require that a uniform speed be maintained to ensure uniform proportioning. Regardless of the type, the manufacturer's recommended procedures for calibrating and operating the travel plant should be followed carefully. Finally, the efficient travel plant should be capable of thoroughly mixing the asphalt and aggregates, uniformly dispersing the asphalt and adequately coating the aggregate particles, thus producing a mixture that is uniform in color.

Hopper travel plants and, in some cases, windrow plants, require devices for ensuring accurate control of the flow of aggregates from the hopper to the pugmill so that correct mix proportions are maintained. Feed of asphalt to the pugmill similarly requires accurate calibration. Typically, a positive displacement pump is utilized to deliver asphalt to the mixing chamber via a spray bar.

4.05 WATER DISTRIBUTOR.—Mixed-in-place construction may require premoistened aggregate for coating and ease of compaction. Consequently, a water distributor, capable of spraying a controlled amount of water, should be available.

4.06 ASPHALT DISTRIBUTOR.—The asphalt distributor is a key piece of equipment in cold mix construction, particularly when rotary pulverizer mixers without built-in asphalt feed are used or when blade mixing is utilized. The asphalt distributor, either truck or trailer-mounted, consists of an insulated tank, self-contained heating system, a pump, and a spray bar and nozzles through which the emulsified or cutback asphalt is applied under pressure onto the prepared aggregate materials (Figure IV-6).

Asphalt distributors range in performance and capability, with some capable of spreading up to 5 m (16.5 ft) wide at controlled rates as high as 13.5 litre/m^2 (3 gal/yd^2).

It is important to keep an adequate supply of asphalt at or near the jobsite to avoid delays. In rural areas, it may be advisable to have an asphalt supply truck at the project.

B. Spreading Equipment

4.07 MOTOR GRADER.—When a motor grader is used to spread asphalt cold mix, it should be checked carefully before putting it into service.

The cutting edge of the blade should be sharp and must be straight from end to end to produce the required cross-section. The blade should be long enough to ensure finishing to close transverse tolerances. Usually a 3.7 or 4 m (12 or 13 ft) blade is used.

The joints and linkage of the blade suspension system should be snug and free from excessive wear. Otherwise, the blade may vibrate or allow irregular pressure during operation and result in surface irregularities. Also, the mold board and circle gear may settle when the machine stops or may climb while spreading.

The motor grader should be heavy enough to hold the blade firmly and uniformly on the surface while spreading the mixture. The wheelbase should be long enough to permit planing to close tolerance. On surface courses, the tires should be

Figure IV-4. Blade Mixing.

Figure IV-5. Hopper-type pugmill travel plant operating from a windrow with a low-lift loader.

Figure IV-6. Asphalt Distributor.

smooth to keep from leaving tread marks in the pavement. The engine should be powerful enough to propel the machine without straining when spreading the mixture.

Automatic controls that hold the blade to a set transverse slope regardless of vertical movements of the grader wheels are available.

4.08 SPREADERS.—Some cold mixes may be spread to the required depth without aeration. Generally, these are open-graded mixes placed under climatic conditions that will allow evaporation of moisture or volatiles within a reasonable time. They may be spread by a rotary mixer, or spread from windrows by motor grader or large multi-purpose equipment. A special pick-up machine can be utilized to place properly windrowed asphalt cold mix into a self-propelled paver. Shown in Figure IV-7 is one piece of equipment that performs pickup and placing in one continuous operation.

C. Compacting Equipment

4.09 PNEUMATIC-TIRED ROLLERS.—Pneumatic-tired rollers used for compacting cold mixes should be the self-propelled tandem type (Figure IV-8). They can have two to seven wheels in front, four to eight in rear, and the wheels should be free to oscillate vertically. They range from 2.7 tonnes (3 tons) empty to 32 tonnes (35 tons) with wet sand ballast.

Figure IV-7. Paver operating from windrow with a pick-up loader.
(Courtesy Barber-Greene Inc.)

Pneumatic-tired rollers are very effective for initial compaction of cold mixes, especially those placed in thin lifts. Final rolling and smoothing of surface defects is best accomplished using steel-wheeled rollers.

A pneumatic-tired roller's weight is but one factor in its compaction effectiveness. Other factors that affect compaction are wheel loads, inflation pressure, contact area of the tire, and rolling speed.

Smooth tires should be used on the roller as treaded tires may leave tread marks that cannot be removed by the final rolling. The contact area of the tire may be adjusted for a given wheel load by increasing the inflation pressure, or by leaving the inflation pressure unchanged and modifying the wheel load. The inflation pressure of the tires usually will be specified with the pressure maintained within ±34 kPa (±5 psi).

4.10 STEEL-WHEELED ROLLERS.—There are two basic types of steel-wheeled rollers.

(1) *Tandem rollers*, two-axle (Figure IV-9), available in weights ranging from 2.7 to 13.6 tonnes (3 to 15 tons) or more. On most, ballast can be added to

Figure IV-8. Pneumatic-tired roller.

Figure IV-9. Steel-wheeled tandem roller.

Figure IV-10. Vibratory roller.

the wheels to increase weight. Although portable tandems are available, with weights in the 2.7 to 5.4 tonne (3 to 6 ton) range, most projects require rollers of 7.2 tonnes (8 tons). Generally, weight exerted by the rear wheel should not be less than 43.8 N/mm (250 lb/in.) of roll width.

(2) *Three-wheel rollers,* equipped with two drive wheels, each usually 1500 to 1800 mm (60 to 70 in.) in diameter by 500 to 600 mm (20 to 24 in.) wide, and a steering wheel, smaller in diameter, but wider. Weights vary from 7.2 tonnes (8 tons) up to 14.5 tonnes (16 tons).

4.11 VIBRATORY ROLLERS.—Vibratory rollers (Figure IV-10) are made with one or two smooth-surfaced steel wheels 900 to 1500 mm (36 to 60 in.) in diameter and 1200 to 2400 mm (48 to 96 in.) in width. These rollers compact by a combination of static weight and dynamic force, with the frequency and amplitude of the force being adjustable. Both the amplitude and frequency, which are affected by vehicle speed, are varied for the particular mix being compacted. In many cases, the best combination is determined on field test sections prior to construction. Vibratory rollers should be equipped with a drum-wetting system and a power hydraulic steering system.

Chapter V. **Construction**

5.01 ROADBED PREPARATION.—The roadbed on which mixed material is placed must be shaped and compacted. The surface must be swept with a power broom to remove dirt and other foreign matter. Depending upon the condition or type of base or subgrade, a prime coat of MC-30, MC-70, or MC-250 may be necessary. When applied, the prime coat is allowed to cure, and after 24 hours any excess (ponding) asphalt remaining on the surface is blotted with sand or other fine material.

5.02 WINDROWS.—Several types of cold mix construction require that the aggregates be placed in windrows prior to mixing and spreading. If windrows are to be used, the roadway must be cleared of all vegetation to a width sufficient to accommodate both windrow and traffic while the mixture cures. Because the thickness of the new pavement is directly proportional to the amount of aggregate in the windrow(s), accurate control and measurement of the volume of the windrowed material is necessary.

Usually, there is not enough loose material on the road surface to use in the road mix. In this case, it is best to blade the loose material onto the shoulder rather than perform the several operations that are necessary to blend it with the material brought in from other sources.

Sometimes, however, incorporating the existing material on the roadbed into the mixture is considered practical, if it is uniform and enough is available. When this is done, the loose aggregate first must be bladed into a windrow and measured. Next, it must be determined if other aggregates are necessary to meet grading specifications. Finally, the windrow is built up to the required volume with imported material to meet the specifications.

If two or more materials are to be combined on the road to be surfaced, each should be placed in its own windrow. These windrows are then mixed together thoroughly before asphalt is added.

5.03 DETERMINING ASPHALT APPLICATION RATE.—Asphalt is either applied with a pressure distributor ahead of the mixing process or, in case of travel mixers, during the mixing process. In any case, close control over the application rate and viscosity is necessary for proper mixing.

Before mixing operations begin, the correct asphalt application rate and forward speed of the spray bar equipped mixer or asphalt distributor must be determined for the quantity of aggregate in the windrow. Also, when using emulsified asphalt, if it is necessary to moisten the aggregate before applying the asphalt, the water application rate and forward speed of the water distributor must be determined.

The following formulas can be used to find the asphalt application rate in litres per linear metre (gallons per linear foot) of windrow and the forward speed required of the mixer or distributor in metres per minute (feet per minute). The formulas also can be used to find the application rate and distributor speed for water by substituting water for asphalt.

Before the correct amount of asphalt to be applied can be determined, it is necessary to determine the quantity of aggregate in the windrow. Measurements to be made on a sized or uniform windrow are shown in Figure V-1. The quantities are then determined by these formulas:

$$V = \frac{(A + B) C}{2}$$

where

V = volume of the windrow, m² per linear metre (ft² per linear foot); and
A, B, C = dimensions of the windrow (Figure V-1), metres (feet)

$$W_f = W_1 V$$

where

W_f = quantity of aggregate, kg per linear metre (lb per linear ft) of windrow; and
W_1 = loose weight of dry aggregate, kg/m³ (lb/ft³)

Example: a windrow of dried aggregate 0.5 m (1.64 ft) high, 0.5 m (1.64 ft) wide at the top, and 1.0 m (3.28 ft) wide at the base has a loose dry weight of 1500 kg/m³ (93.6 lb/ft³). Substituting into the above equations:

Metric Units	Customary Units
$V = \dfrac{(0.5 + 1.0) \, 0.5}{2}$	$V = \dfrac{(1.64 + 3.28) \, 1.64}{2}$
V = 0.375 m² per linear metre	V = 4.03 ft² per linear foot
$W_f = 1500 \times (0.375)$	$W_f = 93.6 \times (4.03)$
W_r = 562.5 kg per linear metre	W_r = 377 lb per linear foot

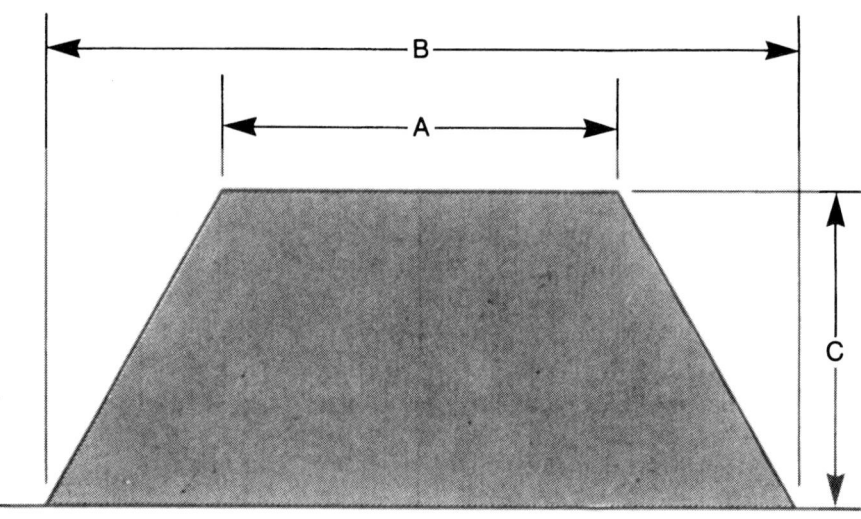

Figure V-1. Measurements for Determining Windrow Quantities.

The rate of application for asphalt along the windrow is determined by:

$$A = \frac{W_f \times P}{100 \times G}$$

where

A = application rate of asphalt, litre per linear metre (gallons per linear foot)
W_f = quantity of aggregate, kg per linear metre (lb per linear foot)
P = asphalt content, percent by weight of dry aggregate; and
G = weight of asphalt, approximately 1 kg/litre (8 lb/gal)

Example:
   Aggregate: weighing 562.5 kg per linear metre (377 lb per linear foot) of windrow
   Asphalt: MC-250, asphalt content = 4 percent by weight of dry aggregate

Metric Units   Customary Units

$$A = \frac{562.5 \times 4}{100 \times 1}$$ $$A = \frac{377 \times 4}{100 \times 8}$$

A = 22.5 litres per linear metre   A = 1.9 gallons per linear foot

To determine forward speed of the distributor:

$$S = \frac{D_p}{A_b}$$

where

S = forward speed of mixer or distributor, $\frac{m}{min}$ $\left[\frac{ft}{min}\right]$

D_p = pump discharge rate, $\frac{litre}{min}$ $\left[\frac{gal}{min}\right]$

A_b = asphalt application rate, $\frac{litre}{m}$ $\left[\frac{gal}{ft}\right]$

Example: A windrow of dry aggregate 0.15 m (0.5 ft) high, 1.5 m (4.9 ft) wide at the top, and 2.0 m (6.6 ft) wide at the base is to be mixed with 5.9 percent by weight of MS-2 emulsified asphalt. The loose unit weight of the aggregate is 1440 kg/m³ (90 lb/ft³). One-half of the asphalt is to be applied in each of two passes of a rotary mixer equipped with a spraying system. Needed is the total asphalt application rate and the forward speed of the mixer.

The first step is determine the volume of aggregate in the windrow. Using the pertinent formulas from the previous page:

$$V = \frac{(A + B)\,C}{2}$$

$$V = \frac{(1.5 + 2.0) \times (0.15)}{2} = 0.26 \text{ m}^3/\text{m} \ (2.8 \text{ ft}^3/\text{ft})$$

And the quantity of aggregate:

$W_f = W_1 V$
$W_f = (1440) \times (.26)$
$W_f = 374.4$ kg/metre (252 lb/ft)

Asphalt application rate:

$$A = \frac{W_f P}{100\,G}$$

$$A = \frac{374.4 \times (5.9)}{100 \times (1)} = 22.1 \text{ litre/m} \ (1.86 \text{ gal/ft})$$

Asphalt application rate per pass $= \frac{22.1}{2} = 11.0$ litre/m (0.93 gal/ft).

Then, the forward speed of the mixer, assuming a constant asphalt pump discharge of 100 litre/min (26.5 gal/min), is

$$S = \frac{100}{11.2} = 9 \text{ m/min } (28.5 \text{ ft/min})$$

5.04 CONTROL OF ASPHALT.—Asphalt is added to the aggregate from an asphalt distributor or by a travel mixer. Whichever method is used, close control of quantity and viscosity is required to ensure a proper mixture.

Proper viscosity is essential, for the asphalt must be fluid enough to flow easily through the spray nozzles and to adequately coat the aggregate particles. Cutback asphalts, although already fluid, need some heating to bring them to the proper viscosity for spraying, which is 20 to 120 centistokes. But unless mixing begins immediately, the viscosity will quickly rise above that recommended for mixing, which is 150 to 300 centistokes. The volatiles in cutback asphalts keep them fluid long enough for the completion of road mixing, if it is done promptly. Even so, the aggregate temperature should be at least 10° C (50° F) in the shade at the time of mixing.

Mixing and spraying temperatures for emulsified and cutback asphalts are given in Table II-2.

5.05 MIXING.—*Rotary Mixing*. Rotary mixers equipped with built-in spraying systems require that the asphalt application rates be matched accurately with the width and thickness of the course, forward speed of the mixer, and the density of the in-place aggregate. However, when utilizing a rotary mixer not equipped with spray bars, an asphalt distributor, operating ahead of the mixer, applies asphalt to the aggregate. Several alternating applications of asphalt and passes of the mixer are usually necessary to achieve the specified mixture.

Most rotary mixers are equipped with a spray system. When using this type of mixer these steps are recommended:

(1) Spread the aggregate to uniform grade and cross section with motor graders.
(2) Thoroughly mix the aggregate with one or more passes of the mixer. (At the time of asphalt application the moisture content of the aggregate should not exceed 3 percent for cutback asphalt, unless laboratory tests indicate that a higher moisture content will not be harmful when the asphalt is added.)
(3) Add asphalt in increments of about 2.25 litres/m^2 (0.50 gal/yd^2) until the total required amount of asphalt is applied and mixed in. A total of 0.7 to 1.1 litre/m^2 per 10 millimeters (0.4 to 0.6 gal/yd^2 per inch) of compacted

(4) Make one or more passes of the mixer between applications of asphalt, as necessary to ensure thorough mixing.
(5) Maintain the surface true to grade and cross-section by using a motor grader during the mixing operations.
(6) Aerate the mixture by additional manipulation, if necessary.

Blade Mixing. With blade mixing, the imported or in-place material is shaped into a windrow. The windrow is then flattened with the blade to about the width of the distributor spray bar. The asphalt is applied by successive passes of the asphalt distributor over the flattened windrow, each application not exceeding 3.5 litre/m² (0.75 gal/yd²).

After each pass of the distributor the mixture is worked back and forth across the roadbed with the blade, sometimes aided by auxiliary mixing equipment. Prior to each succeeding application of asphalt, the mixture is re-formed into a flattened windrow.

The material in the windrow is subjected to as many mixings, spreadings, shapings, and flattenings as are needed to disperse the asphalt thoroughly throughout the mixture and effectively coat the aggregate particles.

During mixing, the vertical angle of the mold board may require adjustment from time to time in order to achieve a complete rolling action of the windrow as it is worked. As large a roll as possible should be carried ahead of the blade, since pressure from the weight of the aggregate facilitates mixing.

Additionally, care must be taken during mixing to see that neither extra material be taken from the mixing table and incorporated into the windrow, nor any of the windrow be lost over the edge of the mixing table or left on the mixing table without being treated.

Sometimes, when cutback asphalt is used, the formation of "oil balls" i.e., concentrated clusters of fine aggregate saturated and coated with excessive amounts of asphalt, can make a mix difficult to spread and compact. This condition can be corrected by forming the mixture into a tight windrow and allowing it to cure for a few days.

After mixing and aeration have been completed, the windrow is moved to one side of the roadbed in readiness for subsequent spreading. If it is left for any length of time, periodic breaks in the windrow should be cut to ensure drainage of rainwater from the roadbed.

Travel-Plant Mixing. Travel-plant mixing offers the advantage of closer control of the mixing operation than is possible with blade mixing.

The hopper-type travel plant operates by mixing, in its pugmill, the proper amount of asphalt with aggregate that is deposited by haul trucks directly into the plant's hopper; it then spreads the mixture. Except when using open-graded mixtures, care must be taken to ensure sufficient evaporation of diluents from the mix prior to compaction.

5.06 AERATION.—Before compaction, most of the diluents that have made the asphalt cold mix workable must be allowed to evaporate. In most cases, this occurs during mixing and spreading and very little additional aeration is required. But, extra manipulation on the roadbed is occasionally needed to help speed the process and dissipate the excess diluents. Until the mix is sufficiently aerated, it usually will not support rollers without excessive shoving. Generally, the mixture is sufficiently aerated when it becomes tacky and appears to "crawl."

Many factors affect the rate and the required amount of aeration. Fine-grained and well-graded mixtures will require longer aeration than open-graded and coarse-grained mixtures, all other things being equal. Also, if an asphalt cold mix base course is to be surfaced within a short length of time, aeration before compaction should be more complete than if the course is not to be surfaced for some time; the surface acts as a seal, greatly retarding the removal of diluents.

Emulsified Asphalt Mixes. Experience has shown that breakdown rolling of emulsified asphalt mixes should begin immediately before, or at the same time as, the emulsion starts to break. (Indicated by a marked color change from brown to black.) About this time, the moisture content of the mixture is sufficient to act as a lubricant between the aggregate particles, but is reduced to the point where it does not fill the void spaces, thus allowing their reduction under compaction forces. Also, by this time, the mixture should be able to support the roller without undue displacement.

Cutback Asphalt Mixes. When using cutback asphalt, correct aeration will be achieved when volatile content is reduced to about 50 percent of that contained in the original asphalt material, and the moisture content is less than 3 percent by weight of the total mixture (refer to ASTM D 1461 or AASHTO T 110).

5.07 SPREADING AND COMPACTING.—With mixing and aeration completed, spreading and compacting the cold mix follows. Achieving a finished section and smooth riding surface conforming to the specifications is the objective of these final two construction steps.

The mixture should always be spread to a uniform thickness (whether in a single pass or in several thinner layers) so that no thin spots exist in the final mat.

Spreading should be accomplished in successive layers, with no layer thinner than about two times the diameter of the maximum particle size nor generally thicker than 76 mm (3 in.). As each layer is spread, compaction should follow almost immediately with a pneumatic-tired roller.

When using a motor grader, there is a tendency for the tires to compact the freshly spread mix. The tracks will appear as ridges in the finished mat unless there is adequate rolling between the spreading of each successive layer. The roller should follow directly behind the motor grader (Figure V-2) in order to eliminate these ridge marks.

If, at any time during compaction, the asphalt mixture exhibits undue rutting or shoving, rolling should be stopped. Compaction should not be attempted until there is a sufficient reduction in moisture or diluent content occurring either naturally or through aeration.

Figure V-2. Spreading and compacting train.

After one course is thoroughly compacted and cured, other courses may be placed on it. This operation should be repeated as many times as necessary to bring the road to proper cross-section. For a smooth riding surface a motor grader can be used to trim and level as the rollers complete compaction of the upper layer.

After the mat has been shaped to its final required cross-section, it must then be finish rolled, preferably with a steel-wheeled roller, until all roller marks are eliminated.

Sometimes, a completed course may have to be opened temporarily to traffic. In this event, to prevent tire pickup, it may be advisable to seal the surface by applying a dilution of slow-setting emulsified asphalt and potable water (in equal parts) at a rate of approximately 0.45 litre/m^2 (0.10 gal/yd^2). This should be allowed to cure until no pickup occurs. For immediate passage of traffic, sanding may be desirable to avoid pickup.

5.08 INSPECTING AND SAMPLING.—The final quality control of materials and construction methods must be accomplished through on-the-job inspection. Sampling and testing must be done as required by the specifications. The responsibility rests with the field personnel to see that the materials used meet the requirements of the specification and that the specified procedures are followed.

Part III: Central Plant Mixing

Chapter VI. Equipment for Plant Mixes

6.01 GENERAL.—Similar to the mixed-in-place method, a variety of equipment is available to carry out plant-mixed cold mix construction. Some of the equipment is common to both methods and, if certain types used for plant-mix construction are discussed in Chapter IV, reference will be made to the applicable article to avoid repetition.

6.02 STATIONARY PLANTS.—Stationary plant mixing generally is accomplished at a location away from the road site, frequently at the aggregate source. A stationary plant consists of a mixer and equipment for heating the asphalt (if necessary) and for feeding the asphalt, aggregate, and additives (if needed) to the mixer. It is similar in many respects to the hot-mix plant, except that it has no dryer or screens other than a scalping screen. Like the hot-mix plant, a stationary cold mix plant may be either a batch (Figure VI-1) or continuous type, although the latter is most prevalently used for cold mix construction (Figures VI-2 and VI-3).

Any type of plant that can produce an asphalt mixture conforming to the specifications can be used. But, as a minimum, it should be equipped with temperature and metering devices to accurately control the asphalt material being applied to the aggregate and controlled feeders for proportioning aggregates and additives. Although not always a plant component, a storage silo allows a more continuous mixing operation, resulting in better mix uniformity.

6.03 HAUL TRUCKS.—Several types of haul trucks may be used for cold mix produced in stationary plants; the type selected depends on the spreading equipment. The traditional raised-bed, end-dump truck can be used with pavers. Bottom dumps produce windrows and are not used with pavers with hoppers unless a low-lift loader is used to transfer the mix to the hopper. Horizontal discharge trucks deposit the mix directly into the paver's hopper without raising the bed. These trucks may also be used with windrow-spreader boxes.

A sufficient number of haul trucks with smooth, clean beds should be available to ensure uniform operation of the mixing plant and paver.

6.04 MOTOR GRADERS.—Motor graders are discussed in Article 4.07.

6.05 PAVERS.—If climatic conditions and aggregate gradation permit evaporation of moisture or volatiles without aeration by manipulation, a conventional self-propelled asphalt paver may be used to place asphalt cold mixture. A full-width paver may be used if the plant can produce enough mixture to keep the paver moving without start-stop operation.

30 Central Plant Mixing

Figure VI-1. Stationary cold mix plant (batch).

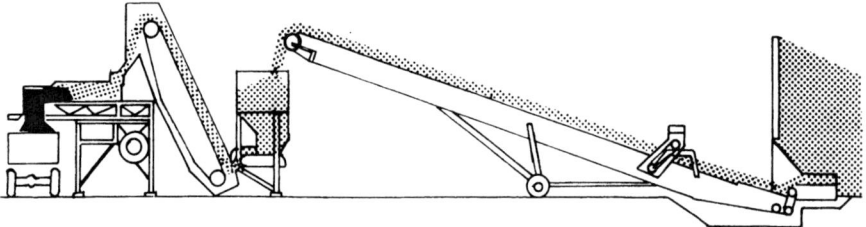

Figure VI-2. Flow diagram of a typical cold mix continuous plant.

Figure VI-3. Cold mix continuous plant.

Asphalt pavers are described in detail in *Principles of Construction of Hot Mix Asphalt Pavements*, (MS-22), Asphalt Institute.

6.06 SPREADERS.—Spreading equipment such as the Jersey spreader and towed spreaders have been commonly used.

A Jersey spreader is a front-wheeled hopper that is attached to the front end of a crawler or rubber-tired tractor. The asphalt mixture is dumped into the hopper and the mixture falls directly to the road where it is spread and struck off to a controlled thickness. To begin spreading the mixture at the specified depth the tractor should be driven onto blocks or boards equal to the depth of the uncompacted spread and placed so that the tractor will ride directly onto the newly-placed material.

Towed-type spreaders are attached to the rear of haul trucks. The asphalt cold mix is deposited into the hopper and falls directly to the surface being paved. As the truck moves forward, the mixture is struck off by a cutter bar, a blade, or by the screed.

The spreader should be towed at a uniform speed for any given setting of the screed or strike-off device. Variations in towing speed will vary spread thickness. Also, frequent stopping and starting may allow settlement of the screed and result in bumps in the pavement surface.

Please refer to Articles 4.9 through 4.11 for discussion of compacting equipment.

Chapter VII. Central Plant Mix Construction

7.01 PREPARATION OF MIXTURES.—In batch-type plants, mixing usually is accomplished by a twin-shafted pugmill. Typically these plants have a capacity of not less than 907 kg (2,000 lb). The correct amounts of asphalt and aggregate, generally determined by weight, are fed into the pugmill. The batch is then mixed and discharged into a haul truck before another batch is produced.

In the continuous-mixing plant, the devices feeding asphalt, aggregate and water, (if needed), are interlocked to automatically maintain the correct proportions. Typically, automatic feeders measure and govern the flow of aggregates in relation to the output of a positive displacement asphalt metering pump. A spray nozzle arrangement at the mixer distributes the asphalt over the aggregate. As the proportioned materials move through the pugmill, completely mixed material, ready for spreading, is discharged for subsequent hauling to the road site (Figure VII-1).

A commonly misunderstood aspect of plant-mixed emulsified asphalt is mixing time. Emulsion mixes usually require shorter mixing time than asphalt hot mixes. The tendency is to over mix emulsified asphalt mixes. The effect is the loss of emulsified asphalt from the coarse aggregate particles and the balling-up of the fine aggregate. It also may result in the premature breaking of the emulsified asphalt, causing overly-stiff mixtures.

A less common problem is insufficient coating of the aggregate, caused by undermixing. Mixing times can be varied in a continuous pugmill plant by changing the arrangement of the paddles, by varying the height of the end gate, or by changing the location of the asphalt spray bar. With a drum-mix plant, mixing time is controlled by varying the slope of the drum or by changing the location of the asphalt inlet pipe within the drum.

It should be noted that with emulsified asphalt, regardless of the mixing method, 100 percent coating of the coarse aggregate particles is not always achieved, neither is it necessary. Additional coating will occur as the mixture is manipulated through the spreading procedure and during rolling. Some aggregate types might be difficult to coat properly, but this fact should be evident at the mix design stage. Mixing procedures should aim at achieving a uniform dispersion of the emulsified asphalt with a complete coating of the finer aggregate fractions.

7.02 AERATING PLANT MIX.—Mixtures that require aeration are generally deposited upon the roadbed in windrows and then spread from these windrows. The cold mix is spread with a motor grader and aerated by blading it back and forth, or it is aerated by rotary tiller mixing equipment.

Refer to Article 5.05 for more information on aeration.

7.03 SPREADING AND COMPACTING.—If mixing moisture can be controlled accurately to a level not requiring aeration or if climatic conditions and aggregate gradation permit evaporation of moisture without aeration by

Figure VII-1. Cold plant mix being loaded into haul truck.

manipulation, a conventional self-propelled asphalt paver may be used to place recycled asphalt cold mix.

Asphalt pavers are described in detail in *Principles of Construction of Hot Mix Asphalt Pavements*, (MS-22), Asphalt Institute.

The successful placement of cold mixes with conventional pavers requires the presence of sufficient fluids. Dry mixes tend to tear beneath the screed or strike-off bar. If the mixture is too dry, the mix water content should be increased. When a self-propelled highway paver is used, heating the screed in an attempt to eliminate this tearing does not help. It actually makes the mix less workable, since it serves to accelerate the drying process.

The mixture should be spread uniformly on the roadbed, beginning at the point farthest from the mixing plant. Hauling over freshly placed material should not be permitted except when required for completion of the work. A light, uniform application of choke aggregate placed before or after the breakdown roller pass will prevent pick-up of the mix by construction traffic or subsequent rolling. The aggregate may consist of coarse, dry sand or the minus 2.00 mm (No. 10) screenings from open-graded aggregate production.

Similar to mixed-in-place, central plant cold mixes gain stability as the moisture or diluents (that have made the mix workable) evaporate. It is important not to hinder this process. Therefore, lift thicknesses are limited by the rate that the mixture loses its moisture or diluent. The most important factors affecting this loss are the type of asphalt, moisture or diluent content, gradation and temperature of the aggregate, wind velocity, ambient temperature, and humidity. Because of these variables, local experience is likely to be the best guide in determining allowable placement thicknesses.

Additional information on spreading and a discussion on compaction are contained in Article 5.07.

7.04 INSPECTING AND SAMPLING.—Please refer to Article 5.08.

Part IV: Appendices

Appendix A: Suggested Plant Mix Guidelines

Guideline PM-1
Cold Asphalt Plant Mix

A. General Requirements

A.01 EQUIPMENT.—The equipment shall include:

(1) One or more asphalt mixing plants designed to produce a uniform mixture within the job-mix tolerances.
(2) One or more self-powered pavers that are capable of spreading the mixture to the thickness and width specified, true to the line, grade and crown shown on the plans.
(3) Enough smooth metal-bedded haul trucks, with covers when required, to ensure orderly and continuous paving operations.
(4) A pressure distributor that is capable of applying tack coat and prime material uniformly without atomization.
(5)* One or more steel-wheeled, pneumatic-tired, or vibratory rollers capable of attaining the required density and smoothness.
(6) A power broom or a power blower or both.
(7) Hand tools necessary to complete the job.

Other equipment may be used in addition to, or in lieu of, the specified equipment when approved by the engineer.

A.02 SAMPLING.—Samples of all materials proposed for use shall be* . . . the engineer for test and analysis.

Sampling of asphalt materials shall be in accordance with the latest revision of AASHTO Designation T40 (ASTM Designation D140). Sampling of mineral aggregate shall be in accordance with the latest revision of AASHTO Designation T 2 (ASTM Designation D 75). Sampling of the asphalt mixture, as required by the engineer, shall be in accordance with the latest revision of AASHTO Designation T 168 (ASTM Designation D 979).

*See "Notes to the Engineer" at end of this Chapter.

A.03 METHODS OF TESTING.—Samples of materials will be tested for the requirements of Section B by the applicable methods specified in this Article. The materials shall not be used until approved by the engineer.

(1) Asphalt materials will be tested by the appropriate AASHTO methods of test designated in Article A.09. If an AASHTO method of test procedure is not available, the equivalent ASTM method will be used.

(2) Mineral aggregates will be tested by one or more of the following AASHTO methods. If an AASHTO method of test procedure is not available, the equivalent ASTM method will be used.

Aggregate Characteristic	Method of Test	
	AASHTO	ASTM
Amount of Material Finer than No. 200 Sieve in Aggregate	T 11	C 117
Unit Weight of Aggregate	T 19	C 29
Sieve Analysis, Fine and Coarse Aggregates	T 27	C 136
Sieve Analysis of Mineral Filler	T 37	D 546
Abrasion of Coarse Aggregate, Los Angeles Machine	T 96	C 131
Plastic Fines in Graded Aggregate and Soils by use of the Sand Equivalent Test	T 176	D 2419

A.04 PLACEMENT LIMITATIONS.—The asphalt paving mixture shall be placed only when the specified density can be obtained. The mixture shall not be placed on any wet surface or when weather conditions will otherwise prevent its proper handling or finishing. Asphalt surface course mixture shall not be placed when the surface temperature of the base course is below 10° C (50° F).

A.05 TRAFFIC CONTROL.—Traffic shall be directed through the project with such signs, barricades, devices, flagmen, and pilot vehicles as necessary to provide maximum safety for the public and the workmen with minimum interruption of the work.

A.06 SAFETY.—Safety precautions shall be used at all times during the progress of the work. As appropriate, workmen shall be furnished with hard hats, safety shoes, respirators, and any other safety apparel that will reduce the possibility of injury from accidents. All Occupational Safety and Health Act requirements shall be observed.

A.07 METHOD OF MEASUREMENT.—The quantities to be paid for will be measured as follows:

(1) *Asphalt Paving Mixture.* Total number of tonnes (tons) of asphalt paving mixture actually used in the work. The mixture shall be measured by truck scales.

(2) *Asphalt for Prime Coat and Tack Coat.* The quantity of asphalt used for prime coat and tack coat will be the litres (gallons) at 15° C (60° F) computed from the quantity measurements in the distributor tank before and after application or the tonnes (tons) measured by truck scales. ASTM Designation D 4311 or the tables in Appendix I can be used to adjust the volumes of cutback asphalt material to the temperature 15° C (60° F). Refer to Table I-7 for temperature corrections of emulsified asphalt.

(3) *Water.* Total number of litres (gallons) of water incorporated into the work.

A.08 BASIS OF PAYMENT.—The quantities measured as described in Article A.07 will be paid for at the contract unit bid price for each item. Payment will be in full compensation for furnishing all materials; for mixing, hauling, and placing the asphalt mixture; for rolling; for use of equipment, tools, and incidentals; and for traffic control necessary to complete the work in accordance with these specifications.

B. Materials

A.09 ASPHALT BINDER.—The type and grade of asphalt material for the paving mixture shall be specified by the engineer prior to the letting of the contract. The specified material shall comply with the applicable requirements of AASHTO Specification M 82, M 140 or M 208 (ASTM Specification D 2027, D 977 or D 2397).

Cutback asphalt for prime coat shall be MC-30, MC-70, or MC-250, complying with the requirements of AASHTO Specification M 82 (ASTM Specification D 2027).

If MC cutback asphalt cannot be used, SS-1, CSS-1, SS-1h, CSS-1h, or MS-2, emulsified asphalt may be used mixed-in prime on the road surface. Emulsified asphalt for prime shall be diluted one part water to one part emulsion and applied at a rate of .5 to 1.4 litre/m^2 (0.1 to 0.3 gal/yd^2) into the top 50 to 75 mm (2 to 3 inches) of scarified base.

Emulsified asphalt for tack coat shall be SS-1, SS-1h, CSS-1, or CSS-1h diluted one part water to one part emulsified asphalt. Before dilution the emulsified asphalt shall comply with the requirements of AASHTO Specification M 140 or M 208 (ASTM Specification D 977 or D 2397).

A.10 MINERAL AGGREGATE.—

(1) *Base Course*. The mineral aggregate for the base course mixture shall be crushed stone, crushed or uncrushed gravel, slag, sand, stone or slag screenings, mineral filler or a combination of two or more of these materials. The combined aggregate shall have a sand equivalent value of not less than 35.*

Slag, if used, shall be air-cooled blast-furnace slag and shall weigh not less than 1.12 tonnes/m^3 (70 lb/ft^3).

Mineral filler shall meet the requirements of ASTM Designation D 242.

(2) *Surface Course*. The mineral aggregate for the surface course mixture shall be crushed stone, crushed gravel, crushed slag, sharp-edged natural sand, mineral filler, or a combination of two or more of these materials. Sixty-five percent by weight of the combined coarse aggregate, other than naturally occurring rough-textured aggregate approved by the engineer, shall consist of crushed pieces having one or more faces produced by fracture.

The combined aggregate shall have a sand equivalent value of not less than 35. Coarse aggregate [material retained on the 2.36 mm (No. 8) sieve] shall have a percent wear by the Los Angeles abrasion machine test of not more than 40 unless specific aggregates having higher values are known to be satisfactory.

Slag, if used, shall be air-cooled blast-furnace slag and shall weigh not less than 1.12 tonnes/m^3 (70 lb/ft^3).

Mineral filler shall meet the requirements of ASTM Designation D 242.

A.11 ASPHALT-AGGREGATE MIXTURE*.—The engineer will

_____ a job-mix formula for each
  (approve)   (specify)

*See "Notes to the Engineer."

mixture. The job-mix formula for the asphalt-aggregate base course mixture shall be within these limits:

Sieve Size*	Total Percent Passing, by Weight	
50 mm (2 in.)	_____	
37.5 mm (1-1/2 in.)	_____	
25.0 mm (1 in.)	_____	
19.0 mm (3/4 in.)	_____	
12.5 mm (1/2 in.)	_____	
9.5 mm (3/8 in.)	_____	
4.75 mm (No. 4)	_____	
2.36 mm (No. 8)	_____	
1.18 mm (No. 16)	_____	
600 μm (No. 30)	_____	
300 μm (No. 50)	_____	
150 μm (No. 100)	_____	
75 μm (No. 200)	_____	
Asphalt Content (Total Liquid)	_____	percent by weight of total mix

This job-mix formula for the asphalt-aggregate surface course mixture shall be within these limits:

Sieve Size*	Total Percent Passing, by Weight	
19.0 mm (3/4 in.)	_____	
12.5 mm (1/2 in.)	_____	
9.5 mm (3/8 in.)	_____	
4.75 mm (No. 4)	_____	
2.36 mm (No. 8)	_____	
1.18 mm (No. 16)	_____	
600 μm (No. 30)	_____	
300 μm (No. 50)	_____	
150 μm (No. 100)	_____	
75 μm (No. 200)	_____	
Asphalt Content (Total Liquid)	_____	percent by weight of total mix

*See "Notes to the Engineer."

The results of single extraction and sieve tests shall not be used as the sole basis of acceptance or rejection of the mixture. Any variation from the job-mix formula in the gradation of the aggregate or in the asphalt content greater than the tolerances shown above shall be investigated and the conditions causing the variation corrected.

The asphalt-aggregate mixture shall meet these test criteria:

Emulsified Asphalt (Appendix E or F)

Stability (Marshall): _____

Stability loss (Marshall): _____ percent

Aggregate Coating: _____ percent

Resistance R_t Value (Hveem): _____

Stabilometer S Value (Hveem): _____

Cohesiometer C Value (Hveem): _____

Cutback Asphalt (Appendix G or H)

Stability (Marshall, Hveem): _____

Solvent Evaporated (Marshall): _____ percent

Stability Retention (Marshall): _____ percent

Flow (Marshall): _____

Moisture Vapor Susceptibility (Hveem): _____

Swell (Hveem): _____ mm (in.)

Air Voids: _____ percent

Voids in Mineral Aggregate: _____ percent

C. Construction

A.12 PREPARING AREA TO BE PAVED.*—

(1) The area to be paved shall be substantially true to line and grade. It shall have a firm and properly prepared surface before paving operations begin. All loose and foreign material shall be removed.

(2) When the compacted subgrade on which the asphalt base is to be placed is loosely bonded, it shall be primed with *. . . litre/m^2 (gal/yd^2) of the type and grade of asphalt priming material designated in Article A.09. The asphalt should be entirely absorbed by the base course and the prime should be fully set and cured before placing the surface.

(3) Holes and depressions in existing surfaces shall be repaired by removing all loose and defective material to sound pavement and replacing with an approved asphalt-aggregate patching material. The patching mixture shall be compacted to produce a tight surface conforming to the adjacent pavement area.

(4) Excess asphalt in patches and joints shall be removed only through methods approved by the engineer.

(5) Immediately prior to application of the asphalt tack coat all loose and foreign material shall be removed by sweeping or by blowing, or both.

(6) Surfaces of curbs, gutters, vertical faces of existing pavements, and all structures to be in actual contact with the asphalt-aggregate mixture shall be given a thin, even coating of asphalt material, type and grade as designated in Article A.09. Care shall be taken to prevent spattering of the asphalt on surfaces that will not be in contact with the asphalt-aggregate mixture.

A.13 TACK COAT.

—If directed by the engineer, a tack coat of *. . . litre/m^2 (gal/yd^2) of diluted emulsified asphalt, of the type and grade designated in Article A.09, shall be applied on each layer of the base course and allowed to cure before placing the succeeding course. The emulsified asphalt shall be diluted with equal parts of water. The tack coat shall be applied on only as much pavement as can be covered with asphalt-aggregate mixture in the same day.

A.14 PREPARING THE MIXTURE.—

(1) The asphalt shall be warmed, if necessary, at the paving plant to a temperature at which it can be applied uniformly to the aggregate.

(2) When it is necessary to blend aggregates from one or more sources to produce the combined gradation, each source or size of aggregate shall be stockpiled individually. Aggregate from the individual stockpiles shall be fed through separate bins to the cold elevator feeders. They shall not be blended in the stockpile.

*See "Notes to the Engineer."

(3) Cold aggregates shall be fed carefully to the plant so that surpluses and shortages will not occur and cause breaks in the continuous operation.
(4) Mixing time shall be the shortest time that will produce a satisfactory mixture.

A.15 PLACING THE MIX.—The base course mixture shall be placed in one or more lifts with an asphalt paver or spreader to provide a nominal compacted thickness of *...mm (in.) The surface course mixtures shall be placed with an asphalt paver to provide a nominal compacted thickness of *....mm (in.) The minimum lift thickness shall be at least two times the maximum particle size. The maximum lift thickness shall be that which can be demonstrated to be laid in a single lift and compacted to a required uniform density and smoothness. Placing the mixture shall be a continuous operation. If any irregularities occur, they shall be corrected before final compaction of the mixture.

A.16 COMPACTING THE MIX.—The mix shall be compacted immediately after placing. Initial rolling with a steel-wheeled tandem or three wheeled roller, vibratory roller, or a pneumatic-tired roller shall follow the paver as closely as possible. If needed, intermediate rolling with a pneumatic-tired roller shall be done immediately behind the initial rolling. An application of choke aggregate may be necessary to prevent mix pick-up by the pneumatic-tired roller. Final rolling shall eliminate marks from previous rolling. In areas too small for the roller a vibrating plate compactor or a hand-tamper shall be used to achieve thorough compaction.

A.17 ACCEPTANCE REQUIREMENTS.*—Divide asphalt mixture production into lots, each lot equal to the mix produced during one day. Determine the target density for each lot by measuring the average density of six laboratory-prepared specimens representing two randomly chosen subsamples from trucks delivering mixture to the jobsite. The target density should be reported as dry density.

Determine the compacted density in the field from five randomly located positions in each lot of the compacted mixture. The density of freshly compacted material can be determined using a properly calibrated nuclear density device or other procedure. Density determinations made after a period of curing may be determined on samples obtained from the compacted material by a suitable core-drilling technique. All compacted densities should be converted to dry density. It is recommended that the average of the five field density determinations made in each lot be equal to or greater than 95 percent of the average density of the six laboratory-prepared specimens, and that no individual determination be lower than 92 percent.

The compacted base and surface shall have average thicknesses no less than those specified on the plans. Any deficiency in thickness shall be made up with surface mixture when the surface course is placed.

*See "Notes to the Engineer."

The surface of the completed pavement will be checked longitudinally and transversely for smoothness with a 3 m (10 ft) straightedge. The surface shall not vary more than 5 mm (0.2 in.) in 3 m (10 ft) parallel to the centerline and not more than 8 mm (0.3 in.) in 3 m (10 ft) at right angles to the centerline.

Notes to the Engineer

Art. A.01 EQUIPMENT, ITEM (5).—A compaction roller can be tandem steel-wheeled, vibratory, or pneumatic-tired. Types not desired should be deleted. For most work, a minimum weight of 9 tonnes (10 tons) and a maximum contact pressure of 620 kPa (90 psi) are recommended.

A finish roller is one used to smooth the mat and iron out imperfections after the mix has been compacted. A tandem steel-wheeled roller with a minimum weight of 7 tonnes (8 tons) is recommended. A finish roller should be required for all jobs.

The number of rollers needed is related to the rate of placement.

Art. A.02 SAMPLES.—First sentence. If the engineer or his representative is to take the samples, the words "taken by" should be inserted. If the contractor is to take the samples, the words "submitted to" should be inserted.

Art. A.10 MINERAL AGGREGATE.—If base course is not to be used, delete text of item (1) and delete "(2)" at the beginning of the second item.

Art. A.11 ASPHALT-AGGREGATE MIXTURE.—Asphalt mix gradations specified by local public agencies may be used if they have a history of satisfactory performance. If a base course is not to be used, delete the second sentence and the base course grading action.

Art. A.11 ASPHALT-AGGREGATE MIXTURE.—*Mixture Gradations.* Delete the sieve sizes that are not needed.

Art. A.11 ASPHALT-AGGREGATE MIXTURE.—*Test Criteria.* Enter the criteria for test limits for each mix in accordance with the method of mix design to be used. In addition, state any special provisions necessary because of the type and quality of local aggregate or because of other local conditions. The test criteria not used should be deleted.

Art. A.12 PREPARING AREA TO BE PAVED.—Delete the paragraphs that are not applicable to the job and renumber the remaining paragraphs.

Art. A.13 TACK COAT.—Depending upon the condition of the old surface, from 0.23 to 0.68 litres/m^2 (0.05 to 0.15 gal/yd^2) of emulsified asphalt, SS-1, SS-1h, CSS-1, or CSS-1h, diluted 1 part water to 1 part emulsified asphalt, usually is sufficient for a tack coat. If a tack coat is not used, the entire article should be deleted.

Art. A.15 PLACING THE MIX.—Insert in the first two sentences the total required thickness of the base course and the surface course. If a base course is not to be placed the first sentence should be deleted.

Art A.17 ACCEPTANCE REQUIREMENTS.—See Appendix D, for random sampling plans.

Art. A.17 ACCEPTANCE REQUIREMENTS.—If a nuclear density device is to be used for measuring the density of the compacted base and surface courses, insert:

> The base and surface will be monitored for density with a nuclear device in accordance with ASTM Method of Test, D 2950.

Art. A.17 ACCEPTANCE REQUIREMENTS.—If a nuclear density device is to be used to determine pavement density, insert:

> The average thicknesses of the base and surface courses will be computed from the average density and weight per square metre (yard) of paving mixture actually used.

Appendix B. **Suggested Mixed-In-Place Guidelines**

Guideline RM-1
Mixed-in-Place Courses

B.01 SCOPE.—Furnish and construct mixed-in-place asphalt courses as specified.

A. General Requirements

B.02 PREPARATION OF ROAD SURFACE.—If any portion of the aggregate for the mixed-in-place course is to be obtained from the existing roadbed, the old road shall be scarified and loosened to the depth shown on the plans. The loosened material shall be thoroughly pulverized, after which all oversize fragments shall be removed from the mixture and discarded.

When all aggregate for the mixed-in-place course is to be brought in as new material, the surface to be covered shall be prepared in accordance with the applicable requirements:

(1) Holes and depressions in granular surfaces shall be repaired by removing all loose and defective material and replacing with granular patching material approved by the engineer. The patching material shall be compacted to produce a tight surface conforming with the adjacent area.
(2) Holes and depressions in old asphalt surfaces shall be repaired by removing all loose and defective material and replacing with an approved asphalt-aggregate patching mixture. The patching mixture shall be compacted to produce a tight surface of the same elevation as the surrounding pavement.
(3) Bumps, waves, and corrugations that impair the riding qualities of the old surface shall be removed to produce a smooth, tight surface.
(4) If an asphalt primer is to be used, (MC-30, MC-70 or MC-250), all loose and foreign material shall be removed by light sweeping and, if it is dusty, the surface shall be dampened with water.
(5) Excess asphalt in patches and joints shall be removed only through methods approved by the engineer. The surface then shall be swept with a rotary broom to remove all loose material. All depressions not reached by the rotary broom shall be cleaned by hand brooming.
(6) Old asphalt surfaces shall be cleaned for the full width to be treated. Dust and other loose material in depressions or other places not reached by mechanical sweepers shall be swept with hand brooms or removed by blowers or flushers.

B.03 EQUIPMENT.—These pieces of equipment shall be used as necessary to complete the specified work: scarifiers; pulverizing equipment; rotary mixers or travel plants; motor graders; windrow devices; aggregate spreaders; power brooms or power blowers; self-propelled vibratory or steel-wheeled tandem and pneumatic-tired rollers capable of attaining the required density; a pressure distributor designed and operated to distribute the asphalt material in a uniform spray without atomization; equipment for heating the asphalt material; a water distributor. Other equipment may be used in addition to, or in lieu of, the specified equipment when approved by the engineer.

B.04 SAMPLES.—Samples of all materials proposed for use shall be submitted by the contractor to the engineer. If any portion of the in-place road materials are to be used in the construction the engineer will furnish the contractor with any test results and improvement requirements for the in-place materials. Samples of all other materials proposed for use under these specifications shall be submitted to the engineer for test and analysis. The material shall not be used until it is approved by the engineer.

Sampling of asphalt materials shall be in accordance with the latest revision of AASHTO Designation T 40 (ASTM Designation D 140). Sampling of mineral aggregate shall be in accordance with the latest revision of AASHTO Designation T 2 (ASTM Designation D 75).

B.05 METHODS OF TESTING.—
(1) Asphalt materials will be tested by the appropriate AASHTO methods of test designated in Article B.11. If an AASHTO method of test is not available, the ASTM method will be used.
(2) Mineral aggregates will be tested, as designated in the detailed requirements of these specifications, by one or more of the following AASHTO methods of test. If an AASHTO method of test is not available, the ASTM method will be used.

Aggregate Characteristic	Method of Test	
	AASHTO	ASTM
Abrasion of Coarse Aggregate, Los Angeles Machine	T 96	C 131
Sieve Analysis, Fine and Coarse Aggregates	T 27	C 136
Unit Weight of Aggregate	T 19	C 29
Sand Equivalent	T 176	D 2419

B.06 WEATHER.—Asphalt shall not be applied to the aggregate when the air temperature in the shade is less than 10° C (50° F) unless otherwise permitted by the engineer. Work shall be suspended during rain or following a rain until the mix dries sufficiently.

B.07 TRAFFIC CONTROL.—Traffic shall be directed through the project with warning signs, flagmen, and pilot trucks or cars in a manner that provides maximum safety for the workmen and traffic and the least interruption of the work.

Traffic shall be kept off freshly sprayed asphalt or mixed materials.

If it is necessary to route traffic over the new work, speed shall be restricted to 40 km/hr (25 miles/hr) or less until rolling is completed and the asphalt mixture is firm enough to take high speed traffic.

B.08 SAFETY.—Safety precautions shall be used at all times during the progress of the work. As appropriate, workmen shall be furnished with hard hats, safety shoes, heat-resistant gloves, respirators, and any other safety apparel that will reduce the possibility of injuries from accidents. All Occupational Safety and Health Act requirements shall be observed.

B.09 METHOD OF MEASUREMENT.—The quantities, as applicable, to be paid for will be:
(1) *Patching Material*—Total number of tonnes (tons) of patching material actually used for patching and reconditioning the base.
(2) *Preparation of Surface*—Total number of square metres (square yards) of surface actually prepared for covering by the asphalt treatment but not including work paid for under item (1).
(3) *Asphalt Materials*—Total number of litres (gallons) at 15° C (60° F) or tonnes (tons) of each asphalt material incorporated into the work. In adjusting volumes of asphalt material to the temperature of 15° C (60° F), ASTM Designation D 4311 or Appendix I will be used as appropriate.
(4) *Mineral Aggregate*—Total number of tonnes (tons) of mineral aggregate incorporated in the work.
(5) *Mixing and Placing*—Total number of square metres (square yards) of road mix laid.
(6) *Water*—Total number of litres (gallons) of water incorporated into the work.

B.10 BASIS OF PAYMENT.—The quantities described in Article B.09 will be paid for at the contract unit price bid for each item. Payment will be in full compensation for furnishing, hauling and placing materials for mixing, for rolling and for all labor and use of equipment, tools, and incidentals necessary to complete the work in accordance with these specifications.

B. Materials

B.11 ASPHALT BINDER.—The asphalt will be specified by the engineer from this table prior to letting the contract.

Asphalt	Specification	
	AASHTO	ASTM
MS-1, MS-2, MS-2h	M 140	D 977
SS-1, SS-1h	M 140	D 977
CMS-2, CMS-2h	M 208	D 2397
CSS-1, CSS-1h	M 208	D 2397
HFMS-1, HFMS-2, HFMS-2h, HFMS-2s	M 140	D 977
MC-250, MC-800, MC-3000	M 82	D 2027
SC-250, SC-800	—	D 2026

The engineer will specify the temperature at which the material shall be used (see Table II-2).

B.12 MINERAL AGGREGATE.—The mineral aggregate for the base course shall be crushed stone, crushed or uncrushed gravel, slag, sand, stone screenings, mineral dust or a combination of any of these materials meeting the gradations and quality requirements of a local public agency.

When the mineral aggregate consists of material in-place in the roadbed all rocks or lumps of material larger than 63 mm (2-1/2 in.) in greatest dimension shall be removed and discarded.

The mineral aggregate for the surface course shall meet the gradation and quality requirements of a local public agency.

C. Construction

Methods of Mixing

B.13 PREPARATION OF MINERAL AGGREGATE.—The total aggregate may be a blend of two or more aggregates which shall result in a combined gradation meeting the specifications for the finished material. The materials may be mixed on the roadbed or on some other approved area off the roadbed by mixing machines or by blade mixing.

When any portion of the mineral aggregate is obtained from the old roadbed it shall be bladed into one or more windrows for measurement and sampling. Where needed, and as directed by the engineer, coarse and fine aggregates shall be brought in and added to the windrowed material. These should be in quantities sufficient to provide a resultant aggregate gradation meeting the specifications and to produce a finished course of specified thickness.

When all aggregate is brought in as new material it shall be deposited in one or more windrows in such quantity and proportions as to provide sufficient total aggregate conforming with the specified gradation and to produce a finished course of the specified thickness.

After the proportions of coarse and fine aggregate are adjusted, the total loose aggregate shall be thoroughly and uniformly mixed. It shall then be placed into one or more truncated windrows of uniform cross-section for final measurement and adjustment.

Alternative No. 1—Blade Mixing

B.14 APPLICATION OF ASPHALT.—When the aggregate material is to be mixed with a motor grader the windrow shall be flattened and the asphalt applied from a distributor. Cutback asphalt shall not be applied when the moisture content of the aggregate exceeds 3 percent, unless laboratory tests indicate that a moisture content in excess of 3 percent at the time the asphalt is added will not be harmful.

The asphalt shall be applied uniformly upon the layer of aggregate at the rate of 2.3 to 4.5 litres/m^2 (0.50 to 1.0 gal/yd^2) at the specified temperature. The initial asphalt application shall then be mixed into the aggregate layer. Successive applications of asphalt shall then be applied and mixed in quantities not exceeding 4.5 litres/m^2 (1.0 gal/yd^2) per application.

B.15 MIXING OPERATION.—Immediately after the first application of asphalt, the aggregate and asphalt shall be thoroughly mixed by motor graders. Mixing shall continue until the asphalt is uniformly distributed over the aggregate. The mixed material shall again be windrowed or spread over the lane being surfaced to a uniform depth, and the second application of asphalt shall be made at the specified rate and in the same manner specified in Article B.14. Mixing shall continue as previously described. These processes shall be repeated until the total amount of asphalt binder specified is mixed with the aggregate and a thoroughly uniform mixture has been obtained.

Alternative No. 2—Travel Mixing

B.16 APPLICATION OF ASPHALT.—Unless the travel mixer is equipped to measure and apply the required amount of asphalt during the mixing operation, such application shall be made directly on the spread aggregate with an asphalt distributor. Cutback asphalt shall not be applied when the moisture content of the aggregate exceeds 3 percent—unless laboratory tests indicate that a moisture content in excess of 3 percent at the time the asphalt is added will not be harmful.

When the travel mixer is equipped to measure and apply asphalt binder, its tanks shall be designed to ensure uniform heating of their entire contents. The contractor shall provide all necessary facilities for determining the temperature of the asphalt during heating and prior to application.

B.17 MIXING OPERATION.—The mixture shall be deposited on the roadway surface either in a windrow at the back of the travel mixer or mechanically spread in a uniform layer so as to produce the specified thickness after compaction. If deposited in a windrow, it shall be spread over the entire roadway surface with approved equipment to the specified thickness and to a smooth, even profile and cross-section.

B.18 AERATION.—Regardless of the mixing method used, manipulation of the mix shall continue until volatiles or water, or both, are removed in quantity sufficient to provide a satisfactory mix.

When mixing and aeration are complete, the mix may be laid and compacted in accordance with Article B.19, or it may be placed in windrows along the edges of the area to be paved for laydown at a later time.

B.19 SPREADING AND COMPACTION.—The mixture shall be uniformly spread over the area to be surfaced to such depths that the material will compact to the specified thickness. This mixture shall be spread for compaction from the large windrow. It shall be bladed from the windrow in a succession of thin layers to a uniform cross-section of specified thickness.

After the mixture is spread as specified, each layer shall be thoroughly and uniformly compacted. Test holes shall be dug at specified intervals to determine the compacted thickness of the layers being placed. Areas which have a deficiency of more than 13 mm (0.5 in.) compacted thickness shall be reworked and enough mixed material added to increase the layer to the depth specified. All irregularities that develop in the surface shall be corrected by blading while the mixture is still soft. Blading and compaction shall continue until the surface is true to grade and cross-section. Final compaction shall be obtained by rolling with a steel-wheeled tandem roller.

Notes to the Engineer

The foregoing guidelines are recommended for use under what may be termed average conditions. It is realized, however, that no single standard specification will satisfactorily cover all variations in local conditions that may prevail for individual jobs. Before adopting these specifications verbatim, the engineer should give particular attention to local conditions and make the changes as necessary.

Prior to letting the contract, the engineer should select the particular type of asphalt materials he wishes to use, deleting the requirements for all other materials shown in these guidelines.

The gradations of mineral aggregates shown in these guidelines are those usually specified in this type of construction. Where aggregates differing from those shown in Article B.12 have provided good pavements, the engineer should substitute the necessary gradation and vary other requirements.

These specifications may call for a finished thickness of pavement ranging from 25 to 300 mm (1 to 12 in.) depending on the purpose for which these materials are being placed. Lifts should be constructed not to exceed 75mm (3 in.) after compaction. However, thicker lifts may be permitted if tests show that specification density can be produced.

For most materials, it has been found that when about 50 percent by weight of the volatiles or water contained in the cutback asphalt have been removed, and the moisture content does not exceed 3 percent by weight of the total mixture, the mix is ready for compaction.

"Standard Method for Moisture or Volatile Distillates in Bituminous Mixtures," AASHTO Designation T 110 (ASTM Designation D 1461), is recommended as a suitable test for determining the percentage of water or volatiles in the mixture.

If it is necessary to import aggregates to adjust the gradation of the material in place on the roadbed, the contractor should be so informed before he is awarded the contract. The gradation of the aggregate that must be imported, and the locations where adjustments are necessary, should be generally indicated either in the specifications or on the plans—in order that the contractor will be in a position to make an intelligent bid.

A seal coat or hot mix surface should not be placed until the road mix is thoroughly cured.

Guideline RM-2
Road-Mixed Asphalt Courses for Base and Surface (Sand or Soil)

B.20 SCOPE.—Furnish and construct mixed-in-place sand- or soil-asphalt courses as specified.

A. General Requirements

B.21 PREPARATION OF ROADWAY.—This type of construction may be performed either with the natural material occurring in the roadbed or by importing materials from pits or other sources.

When the material to be treated is that already in the roadbed, it shall be scarified with approved equipment to a depth 50 mm (2 in.) greater than the specified depth of pavement to be constructed and to a width 0.6 m (2 ft) outside the proposed edge of the pavement. After the material has been scarified it shall be thoroughly mixed and processed. All foreign substances shall be removed and discarded. Any particles of aggregate that will not pass a 63 mm (2-1/2 in.) square opening screen shall be removed and discarded. If needed, imported material meeting the requirements of Article B.31 shall be thoroughly mixed with the in-place material.

B.22 EQUIPMENT.—These pieces of equipment shall be used as necessary to complete the specified work: scarifiers; pulverizing equipment; rotary mixers or travel plants; motor graders; windrow devices; aggregate spreaders; power brooms or power blowers; self-propelled vibratory or steel-wheeled tandem and pneumatic-tired rollers capable of attaining the required density; a pressure distributor designed and operated to distribute the asphalt material in a uniform spray without atomization; equipment for heating the asphalt material; a water distributor. Other equipment may be used in addition to, or in lieu of, the specified equipment when approved by the engineer.

B.23 SAMPLES.—Samples of all materials proposed for use shall be submitted by the contractor to the engineer. If any portion of the in-place road materials are to be used in the construction the engineer will furnish the contractor with any test results and improvement requirements for the in-place materials. Samples of all other materials proposed for use under these specifications shall be submitted to the engineer for test and analysis. The material shall not be used until it is approved by the engineer.

Sampling of asphalt materials shall be in accordance with the latest revision of AASHTO Designation T 40 (ASTM Designation D 140). Sampling of mineral aggregate shall be in accordance with the latest revision of AASHTO Designation T 2 (ASTM Designation D 75).

B.24 METHODS OF TESTING.—
(1) Asphalt materials will be tested by the appropriate AASHTO method of test designated in Article B.30. If an AASHTO method of test is not available, the ASTM method will be used.
(2) Mineral aggregates will be tested by one or more of the following AASHTO methods of test. If an AASHTO method of test is not available, the ASTM method will be used.

	Method of Test	
Aggregate Characteristic	AASHTO	ASTM
Sieve Analysis, Fine and Coarse Aggregates	T 27	C 136
Unit Weight of Aggregate	T 19	C 29
Sand Equivalent	T 176	D 2419
Plasticity Index of Soils	T 90	D 424

B.25 WEATHER.—Asphalt shall not be applied to the aggregate when the air temperature in the shade is less than 10° C (50° F) unless otherwise permitted by the engineer. Work shall be suspended during rain or following the rain until mix has sufficiently dried.

B.26 TRAFFIC CONTROL.—Traffic shall be directed through the project with warning signs, flagmen, and pilot trucks or cars in a manner that provides maximum safety for the workmen and traffic and the least interruption of the work.

Traffic shall be kept off freshly sprayed asphalt or mixed materials.

If it is necessary to route traffic over the new work, speed shall be restricted to 40 km/hr (25 miles/hr) or less until rolling is completed and the asphalt mixture is firm enough to take high speed traffic.

B.27 SAFETY.—Safety precautions shall be used at all times during the progress of the work. As appropriate, workmen shall be furnished with hard hats, safety shoes, gloves, respirators, and any other safety apparel that will reduce the possibility of accidents. All Occupational Safety and Health Act requirements shall be observed.

B.28 METHOD OF MEASUREMENT.—The quantities, as applicable, to be paid for will be:
  (1) *Preparation of Surface*—Total number of square metres (square yards) of surface actually prepared for covering by the asphalt treatment.
  (2) *Asphalt Materials*—Total number of litres (gallons) at 15° C (60° F) or tonnes (tons) of each asphalt material incorporated into the work. In adjusting volumes of asphalt material to the temperature of 15° C (60° F), ASTM Designation D 4311 or Appendix I will be used as appropriate.
  (3) *Mineral Aggregate*—Total number of tonnes (tons) of mineral aggregate incorporated in the work.
  (4) *Mixing and Placing*—Total number of square metres (square yards) of road mix laid.
  (5) *Water*—Total number of litres (gallons) of water incorporated into the work.

B.29 BASIS OF PAYMENT.—The quantities described in Article B.28 will be paid for at the contract unit price bid for each item. Payment will be in full compensation for furnishing, hauling and placing materials for mixing, for rolling and for all labor and use of equipment, tools, and incidentals necessary to complete the work in accordance with these specifications.

B. Materials

B.30 ASPHALT BINDER.—The asphalt will be specified by the engineer from this table prior to letting the contract.

Asphalt	AASHTO Specs.	ASTM Specs.
SS-1, SS-1h	M 140	D 977
CSS-1, CSS-1h	M 208	D 2397
HFMS-2s	M 140	D 977
MC-70, MC-250, MC-800	M 82	D 2027

The engineer will specify the temperature at which the material shall be used (see Table II-2).

B.31 MINERAL AGGREGATE.—The mineral aggregate shall consist of material naturally occurring in the roadbed; material imported from local pits or other sources, with or without mineral filler; or any combination of these aggregates that will meet these requirements:
  (1) Passing a 75 µm (No. 200) sieve, not more than 25 percent.
  (2) Sand Equivalent, not less than 35, or Plasticity Index, not more than 6.

C. Construction

Methods of Mixing

B.32 PREPARATION OF MINERAL AGGREGATE.—Where mixing of the aggregate is to be done by means other than a travel mixer, any mineral filler or other aggregate to be blended with the natural material shall be spread over the surface of the scarified material in a uniform quantity, and in such quantity that will provide a mixture meeting the requirements of Article B.31. Such applications shall be made immediately after the scarifying operations; mixing with a rotary-type mixer shall continue until a uniform mixture is obtained.

Where a travel mixer is to be used, the prepared in-place material shall be bladed into one or more windrows suitable for the type of travel mixer. Any additional aggregate required to be blended with the windrowed material shall be uniformly distributed over the windrows as directed by the engineer. Windrows shall contain sufficient material to produce the required thickness of compacted pavement.

If all aggregate material is to be imported from local pits or other sources, this shall be spread on the prepared subgrade or placed in windrows (depending on the method of mixing that will be used) in quantities sufficient to produce the required pavement thickness.

Alternative No. 1—Blade Mixing

B.33 APPLICATION OF ASPHALT.—When the aggregate material is to be mixed with a motor grader, the windrow shall be flattened and the asphalt applied with a distributor. Cutback asphalt shall not be applied when the moisture content of the aggregate exceeds 3 percent, unless laboratory tests indicate that a moisture content in excess of 3 percent at the time the asphalt is added will not be harmful.

The asphalt shall be applied uniformly upon the layer of aggregate material at the rate of 2.3 to 4.5 litres/m^2 (0.50 to 1.0 gal/yd^2) at the specified temperature. The initial asphalt application shall then be mixed into the layer. Successive applications of asphalt shall then be applied and mixed in quantities not exceeding 4.5 litres/m^2 (1.0 gal/yd^2).

B.34 MIXING OPERATION.—As soon as the total specified amount of asphalt is applied to the aggregate material, mixing shall be continued with motor graders until a thoroughly uniform mixture is produced.

Alternative No. 2—Travel Mixing

B.35 APPLICATION OF ASPHALT.—Cutback asphalt shall not be applied when the moisture content of the aggregate material exceeds 3 percent, unless laboratory tests indicate that a moisture content in excess of 3 percent at the time the asphalt is added will not be harmful. If the mixer is not equipped to measure and apply the asphalt during the mixing operation, the asphalt shall be applied directly on the measured windrows with the asphalt distributor. When the mixer is equipped to measure and apply the asphalt, the application will be made during the mixing process.

B.36 MIXING OPERATION.—The aggregate material and asphalt shall be thoroughly mixed until the asphalt is uniformly distributed throughout. The mixture shall be placed in a windrow for later spreading, aeration, and compaction.

B.37 AERATION.—Regardless of the mixing method used, manipulation of the mix shall continue until volatiles or water, or both, are removed in a quantity sufficient to provide a satisfactory mix.

When mixing and aerating are complete, the mix may be laid and compacted in accordance with Article B.38, or it may be placed in windrows along the edges of the area to be paved for laydown at a later time.

B.38 SPREADING AND COMPACTION.—After the material has been aerated it shall be spread to a uniform grade and cross-section and compacted with a pneumatic-tired roller for the full width of the roadway. Rolling shall continue until the entire depth is compacted to the specified density. Test holes shall be dug at specified intervals to determine the compacted thickness of the layers being placed. Areas which have a deficiency of more than 13 mm (0.5 in.) from the specified compacted thickness shall be reworked with enough additional mixed material to increase the layer to the depth specified. All irregularities that develop in the surface shall be corrected by blading while the pavement is still soft. Blading and compaction shall continue until the surface is true to grade and cross-section.

B.39 APPLICATION OF SEAL COAT.—Upon the thoroughly-cured asphalt course, emulsified asphalt, (RS-1, RS-2, CRS-1, or CRS-2,) shall be uniformly applied and immediately covered with aggregate. The procedure for this operation shall be in accordance with Asphalt Surface Treatments–Specifications (ES-11) and Asphalt Surface Treatments–Construction Techniques (ES-12), Asphalt Institute.

Notes to the Engineer

The foregoing guidelines for road-mixed asphalt courses are recommended for use under what may be termed average conditions. It is realized, however, that no single standard specification will satisfactorily cover all variations in local conditions which may prevail for individual jobs. Before adopting these guidelines verbatim the engineer, therefore, should give particular attention to the items listed below and, if necessary, make the changes suggested.

Asphalt-sand and asphalt-soil mixed-in-place courses are usually laid to a compacted thickness from 75 to 150 mm (3 to 6 in.) depending upon traffic conditions. However, greater thicknesses dictated by local conditions of soil/aggregate characteristics may sometimes be allowable.

Prior to letting the contract the engineer should select the particular asphalt material he wishes to use, deleting the requirements for all other asphalt materials shown in these guidelines.

Asphalt-sand or soil mixes normally serve better as base courses. But in some localities, because of lack of aggregate and in the interest of economy, they may be used as surface courses.

The loose grading requirement in Article B.31 is included in this guideline to allow the use of local sands and soils that may vary widely in grading but are still suitable for mixing with asphalt.

Appendix C. Suggested Guidelines for Stockpile Patching Mixtures

Guideline PM-2
Plant-Mixed Asphalt Stockpile Maintenance Mixtures

C.01 SCOPE.—Furnish stockpile asphalt maintenance mixture as specified.

A. General Requirements

C.02 EQUIPMENT.—The equipment shall include an asphalt mixing plant designed, coordinated, and operated to produce a uniform mixture within the job-mix tolerances.

C.03 SAMPLES.—Samples for all materials proposed for use under these specifications shall be submitted to the engineer for test and analysis. The material shall not be used until it is approved by the engineer.

Sampling of asphalt materials shall be in accordance with the latest revision of AASHTO Designation T40 (ASTM Designation D140). Sampling of mineral aggregate shall be in accordance with the latest revision of AASHTO Designation T 2 (ASTM Designation D 75).

C.04 METHODS OF TESTING.—
(1) Asphalt materials will be tested by the appropriate methods of test designated in Article C.05. If an AASHTO method of test is not available, the ASTM method will be used.
(2) Mineral aggregates will be tested by one or more of the following AASHTO methods of test. If an AASHTO method of test is not available, the ASTM method will be used.

	Method of Test	
Aggregate Characteristic	AASHTO	ASTM
Abrasion of Coarse Aggregate, Los Angeles Machine	T 96	C 131
Sieve Analysis, Fine and Coarse Aggregates	T 27	C 136
Unit Weight of Aggregate	T 19	C 29
Sand Equivalent	T 176	D 2419

B. Materials

C.05 ASPHALT BINDER.—The asphalt will be specified by the engineer from this table prior to letting the contract.

Asphalt	AASHTO Specs.	ASTM Specs.
MC-250, MC-800	M 82	D 2027
SC-250, SC-800	—	D 2026
HFMS-2s	M 140	D 977
CMS-2, CMS-2h	M 208	D 2397

The engineer will specify the temperature at which the material shall be used (see Table II-2).

C.06 MINERAL AGGREGATE.—The mineral aggregate shall be crushed stone, crushed or uncrushed gravel, slag, sand, stone or slag screenings, mineral dust or a combination of any of these materials meeting one of these gradations:

Sieve Sizes	Gradation 1	Gradation 2	Gradation 3
25.0 mm (1 in.)	—	—	100
19.0 mm (3/4 in.)	—	100	90-100
12.5 mm (1/2 in.)	100	90-100	—
9.5 mm (3/8 in.)	90-100	—	60-80
4.75 mm (No. 4)	60-80	45-70	35-65
2.36 mm (No. 8)	35-65	25-55	20-50
300 µm (No. 50)	6-25	5-20	3-20
75 µm (No. 200)	2-10	2-9	2-8

The combined aggregate shall have a sand equivalent value of not less than 35. Coarse aggregate [material retained on the 2.36 mm (No. 8) sieve] shall have a percent wear by the Los Angeles abrasion machine test of not more than 50, unless specific aggregates having higher values are known to be satisfactory.

C. Mixture Preparation

C.07 PREPARATION OF MIXTURES.—Coarse and fine aggregate shall be fed into the plant in the proportions required to provide a total aggregate conforming with the grading specified in Article C.06. The aggregate used with cutback asphalt shall have a moisture content less than 3 percent at the time of mixing unless tests indicate a higher moisture content will not be detrimental. The asphalt shall be applied at the rate and the temperature specified by the engineer. The mineral aggregate and asphalt shall be mixed thoroughly until all aggregate particles are completely coated.

C.08 STOCKPILING.—The completed mixture shall be stored in a clean area to prevent contamination. A covered storage bin will protect it and will help retain workability.

C.09 METHOD OF MEASUREMENT.—The quantities to be paid for will be the total number of tonnes (tons) of asphalt maintenance mixture delivered to the stockpile.

C.10 BASIS OF PAYMENT.—The quantities measured as described in Article C.09 will be paid for at the contract unit price bid for this item. Payment will be in full compensation for furnishing, mixing, hauling, and stockpiling the mixture and for all labor and use of equipment, tools, and incidentals necessary to complete the work in accordance with these specifications.

Notes to the Engineer

(1) *Selection of Asphalt*—This guide may be used for selecting the type and grade of asphalt for the stockpile mixture:

MC-250—For immediate use under hot or moderate weather conditions, or otherwise for use within a short time after stockpiling.

MC-800—For use within short time after stockpiling.

SC-250—For long period storage in hot, dry climates.

SC-800—For long period storage.

CMS-2—Mix can be designed for use within a short time after stockpiling.

CMS-2h—Mix can be designed for use within a short time after stockpiling.

HFMS-2s—Mix can be designed for use within a medium to long time after stockpiling.

(2) *Amount of Asphalt*—The amount of cutback asphalt required for the aggregate grading specified in Article C.06 will normally be in the range of 4 to 6 percent by weight of total mix; for emulsified asphalt the requirement is typically 7 to 10 percent.

(3) It should be recognized that due to the higher asphalt content and lower unit weights of some stockpile mixtures, the plant may be required to operate outside of the normal range for mixing. Special attention should be made to the control and monitoring of weights, flows and temperatures.

(4) Aggregates with high absorption may be unsuitable for stockpile mixtures. The high solvent content and normal low viscosity of many asphalt materials for stockpile mixtures may result in a high penetration of asphalt into the aggregate and cause a dry mix.

Guideline RM-3
Mixed-in-Place Asphalt Stockpile
Maintenance Mixtures

C.11 SCOPE.—Furnish stockpile asphalt maintenance mixture as specified.

A. General Requirements

C.12 EQUIPMENT.—The equipment includes travel mixers; rotary mixers; motor graders; and an asphalt pressure distributor meeting these requirements:

The pressure distributor shall be designed and operated to distribute the asphalt material in a uniform spray without atomization.

The distributor shall be equipped with a bitumeter having a dial registering metres (feet) of travel per minute. The dial shall be visible to the truck driver so that he can maintain the constant speed required for application at the specified rate.

The pump shall be equipped with a tachometer having a dial registering litres (gallons) per minute passing through the nozzles. The dial shall be readily visible to the operator.

Means for accurately indicating the temperature of the asphalt material in the distributor at all times shall be provided. The thermometer shall not be in contact with, or close to, a heating tube.

The normal width of application of the spray bar shall not be less than 3.5 m (12 ft) with provision for the application of lesser widths when necessary.

The distributor shall be equipped with heating attachments and the asphalt material shall be circulated through the spray bar during the entire heating process.

C.13 SAMPLES.—Samples of all materials proposed for use under these specifications shall be submitted to the engineer for test and analysis. The material shall not be used until it is approved by the engineer.

Sampling of asphalt materials shall be in accordance with the latest revision of AASHTO Designation T40 (ASTM Designation D140). Sampling of mineral aggregate shall be in accordance with the latest revision of AASHTO Designation T 2 (ASTM Designation D75).

C.14 METHODS OF TESTING.—
(1) Asphalt materials will be tested by the appropriate AASHTO methods of test designated in Article C.15. If an AASHTO method of test is not available, the ASTM method will be used.
(2) Mineral aggregates will be tested by one or more of the following AASHTO methods of test. If an AASHTO method of test is not available, the ASTM method will be used.

	Method of Test	
Aggregate Characteristic	AASHTO	ASTM
Abrasion of Coarse Aggregate, Los Angeles Machine	T 96	C 131
Sieve Analysis, Fine and Coarse Aggregates	T 27	C 136
Unit Weight of Aggregate	T 19	C 29
Sand Equivalent	T 176	D 2419

B. Materials

C.15 ASPHALT BINDER.—The asphalt will be specified by the engineer from this table prior to letting the contract.

Asphalt	AASHTO Specs.	ASTM Specs.
MC-250, MC-800	M 82	D 2027
SC-250, SC-800	—	D 2026
HFMS-2s	M 140	D 977
CMS-2, CMS-2h	M 208	D 2397

The engineer will specify the temperature at which the material shall be used (see Table II-2).

C.16 MINERAL AGGREGATE.—The mineral aggregate shall be crushed stone, crushed or uncrushed gravel, slag, sand, stone screenings, mineral dust or a combination of any of these materials meeting one of these gradations:

Sieve Sizes	Gradation 1	Gradation 2	Gradation 3
25.0 mm (1 in.)	—	—	100
19.0 mm (3/4 in.)	—	100	90-100
12.5 mm (1/2 in.)	100	90-100	—
9.5 mm (3/8 in.)	90-100	—	60-80
4.75 mm (No. 4)	60-80	45-70	35-65
2.36 mm (No. 8)	35-65	25-55	20-50
300 µm (No. 50)	6-25	5-20	3-20
75 µm (No. 200)	2-10	2-9	2-8

The combined aggregate shall have a sand equivalent value of not less than 35. Coarse aggregate [material retained on the 2.36 mm (No. 8) sieve] shall have a percent wear by the Los Angeles abrasion machine test of not more than 50, unless specific aggregates having higher values are known to be satisfactory.

C. Construction

Alternative No. 1—Blade Mixing

C.17 PREPARATION OF MIXTURE.—

(1) Coarse and fine mineral aggregate shall be deposited in a single windrow in the proportions required to provide a total aggregate conforming with the gradation specified in Article C.16. After the proportions of coarse and fine aggregate are adjusted, the total loose aggregate shall be mixed thoroughly with a motor grader. It shall then be bladed into a single windrow of uniform cross-section for measurement and adjustment as directed by the engineer.

(2) Immediately prior to application of the asphalt, the windrow of mixed aggregate shall be bladed to a uniform cross-section approximately 50 mm (2 in.) thick. It shall be bladed back and forth to reduce moisture content, if necessary. Upon the layer of graded aggregate, the asphalt shall be uniformly applied with the asphalt distributor at the rate of 2.3 to 4.5 litres/m^2 (0.5 to 1.0 gal/yd^2) at the specified temperature. The aggregate and asphalt shall be mixed as described in paragraph (3). Successive treatments of asphalt shall then be applied and mixed in the quantities, not exceeding 4.5 litres/m^2 (1.0 gal/yd^2) each, as directed by the engineer.

(3) The motor grader shall follow the distributor immediately after each application of asphalt, and shall continue to operate on the treated strip until all asphalt is mixed into the aggregate. After the aggregate has received its total application of asphalt, mixing shall continue until a thoroughly uniform mixture is produced. If, before the process is completed, the mixture should become wet, the mixing operation shall be continued until it dries out. After final mixing the material shall be brought to a single windrow.

Alternative No. 2—Travel Mixing

C.18 PREPARATION OF MIXTURE.—Coarse and fine aggregate shall be deposited in a single windrow in the proportions required to provide a total aggregate conforming with the gradation specified in Article C.16. Unless the travel mixer is equipped to measure and apply asphalt during the mixing operation, the windrow shall be flattened and the application shall be made with the asphalt distributor. The aggregate shall be free from visible moisture at the time of mixing. The asphalt shall be applied at the temperature and rate specified by the engineer. The mineral aggregate and the asphalt shall be mixed thoroughly until all aggregate particles are completely coated.

C.19 STOCKPILING.—The completed mixture shall be stored in a clean area to prevent contamination. A covered storage bin will protect it and will help retain workability.

C.20 METHOD OF MEASUREMENT.—The quantities to be paid for will be the total number of tonnes (tons) of asphalt maintenance mixture delivered.

C.21 BASIS OF PAYMENT.—The quantities measured as described in Article C.20 will be paid for at the contract unit price bid for this item. Payment will be in full compensation for furnishing, mixing, hauling, and stockpiling the mixture and for all labor and use of equipment, tools, and incidentals necessary to complete the work in accordance with these specifications.

Notes to the Engineer

(1) *Selection of Asphalt*—This guide may be used for selecting the type and grade of asphalt for the stockpile mixture:

MC-250—For immediate use under hot or moderate weather conditions, or otherwise for use within short time after stockpiling.

MC-800—For use within short time after stockpiling.

SC-250—For long period storage in hot, dry climates.

SC-800—For long period storage.

CMS-2—Mix can be designed for use within a short time after stockpiling.

CMS-2h—Mix can be designed for use within a short time after stockpiling.

HFMS-2s—Mix can be designed for use within a medium to long time after stockpiling.

(2) *Amount of Asphalt*—The amount of cutback asphalt required for the aggregate grading specified in Article C.16 will normally be in the range of 4 to 6 percent by weight of total mix; for emulsified asphalt, the requirement is typically 7 to 10 percent.

(3) Aggregate with high absorption may be unsuitable for stockpile mixtures. The high solvent content and normal low viscosity of many asphalt materials for stockpile mixtures may result in a high penetration of asphalt into the aggregate and cause a dry mix.

Appendix D. Random Sampling Plans

D.01 SELECTING SAMPLING LOCATIONS IN TRUCKS HAULING ASPHALT MIXTURE.—These definitions apply (see also Figure D-1):

- *Lot*—a quantity of material that one desires to control. It may represent a day's production, a specified tonnage, a specified number of truckloads, a specified time period during production.
- *Sample*—a segment of a lot chosen to represent the total lot. It may represent any number of subsamples.
- *Subsample*—a segment of a sample taken from a unit of the lot, i.e., a specified ton, a specified time, a specified truckload.
- *Sample Unit*—a portion of a subsample taken from a unit of a lot and combined with one or more other sample units to make up a subsample.

In this procedure these steps are necessary to select the sampling locations:

(1) Select the lot size—it can be time (hours), an average day's production (tonnes), a selected tonnage (example: 2,000 tonnes), or a selected number of truckloads. (A lot size equal to a day's production is recommended for this procedure as being convenient and easy to randomize.)

(2) Select the number of samples desired per lot. One sample per lot, made up of *four* subsamples, is the minimum recommended.

(3) Select the number of locations in each truckload from which sample units of asphalt mixtures will be taken to combine into one subsample. *Two* sample units per subsample are recommended.

(4) Assign each truckload of mixture in the lot a number, beginning with 1 for the first truckload and number them successively to the highest number in the lot. Find the truckload numbers for sampling by this procedure:
- Place consecutively numbered 25 mm (1 in.) square pieces of cardboard equal to the number of truckloads in the lot into a container (such as a bowl). Mix them thoroughly before each drawing.
- Draw a number of cardboard squares from the container equal to the number of subsamples desired for the lot. The numerals on the cardboard squares will be the truckloads to be sampled.

(5) Choose for each subsample desired the location in the truckload for each of the sample units:
- Divide the truck beds into equal quadrants and number them 1 through 4 in any order desired.
- Place four consecutively numbered (1 through 4) 25 mm (1 in.) square pieces of cardboard into a container (such as a bowl). Mix them thoroughly before each drawing.
- Draw out an amount of cardboard squares equal to the number of sample units desired. The numerals on each square drawn represent the quadrants from which the sample unit will be taken. Replace the cardboard squares and repeat this step for each sample unit of each subsample to be taken.

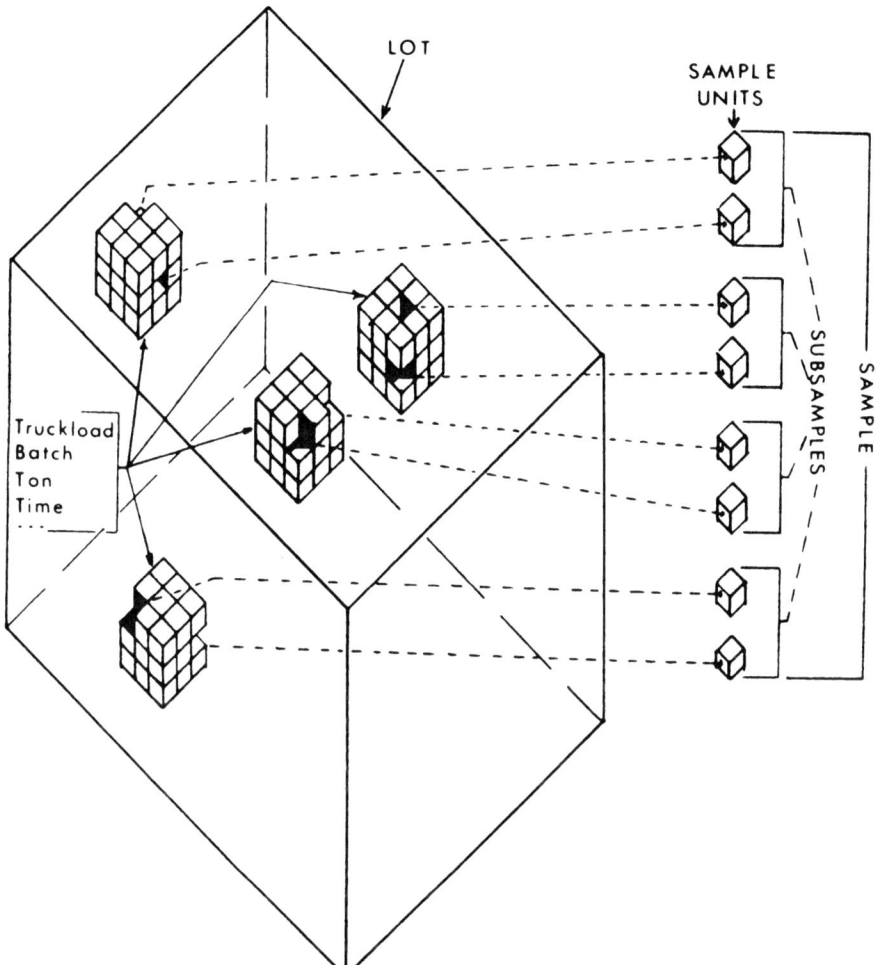

Figure D-1. Schematic diagram illustrating Lot, Sample, Subsample, and Sample Unit.

D.02 SELECTING SAMPLING LOCATIONS AT THE PAVEMENT
SITE.—Table D-1 contains random numbers for the general sampling procedure. To use this table for selecting locations for sampling or testing, these steps are necessary:
(1) *For compacted pavement sampling or testing locations,* use each day's run as a separate section.
(2) Determine the number of sampling locations within a section by selecting the maximum *average* longitudinal distance desired between samples and dividing the length of the section by the maximum average longitudinal distance.
(3) Select a column of random numbers in Table D-1 by placing 28 consecutively numbered pieces of 25 mm (1 in.) square cardboard into a container (such as a bowl), shaking them to get them thoroughly mixed, and drawing one out.
(4) Go to the column of random numbers identified with the number drawn from the container. In subcolumn A, locate all numbers equal to and less than the number of sampling locations per section desired.
(5) Multiply the total length of the section by the decimal values in subcolumn B, found opposite the numbers located in subcolumn A. Add the result to the station number at the beginning of the section to obtain the station of the sampling location.
(6) Multiply the total width of the proposed pavement in the section by the decimal values in subcolumn C, found opposite the numbers located in subcolumn A, then subtract one-half of the total width of the proposed pavement from the result to obtain the offset distance from the centerline to the sampling location. A positive (+) number will be the distance to the *right* of centerline and a negative (-) number will be the distance to the *left* of centerline. If only one lane of pavement is involved, the total width will be the lane width and the offset distance will be measured from the left edge of the lane.

Table D-1. Random Numbers for General Sampling Procedure

Col. No. 1			Col. No. 2			Col. No. 3			Col. No. 4			Col. No. 5			Col. No. 6			Col. No. 7		
A	B	C	A	B	C	A	B	C	A	B	C	A	B	C	A	B	C	A	B	C
15	.033	.576	05	.048	.879	21	.013	.220	18	.089	.716	17	.024	.863	30	.030	.901	12	.029	.386
21	.101	.300	17	.074	.156	30	.036	.853	10	.102	.330	24	.060	.032	21	.096	.198	18	.112	.284
23	.129	.916	18	.102	.191	10	.052	.746	14	.111	.925	26	.074	.639	10	.100	.161	20	.114	.848
30	.158	.434	06	.105	.257	25	.061	.954	28	.127	.840	07	.167	.512	29	.133	.388	03	.121	.656
24	.177	.397	28	.179	.447	29	.062	.507	24	.132	.271	28	.194	.776	24	.138	.062	13	.178	.640
11	.202	.271	26	.187	.844	18	.087	.887	19	.285	.899	03	.219	.166	20	.168	.564	22	.209	.421
16	.204	.012	04	.188	.482	24	.105	.849	01	.326	.037	29	.264	.284	22	.232	.953	16	.221	.311
08	.208	.418	02	.208	.577	07	.139	.159	30	.334	.938	11	.282	.262	14	.259	.217	29	.235	.356
19	.211	.798	03	.214	.402	01	.175	.641	22	.405	.295	14	.379	.994	01	.275	.195	28	.264	.941
29	.233	.070	07	.245	.080	23	.196	.873	05	.421	.282	13	.394	.405	06	.277	.475	11	.287	.199
07	.260	.073	15	.248	.831	26	.240	.981	13	.451	.212	06	.410	.157	02	.296	.497	02	.336	.992
17	.262	.308	29	.261	.087	14	.255	.374	02	.461	.023	15	.438	.700	26	.311	.144	15	.393	.488
25	.271	.180	30	.302	.883	06	.310	.043	06	.487	.539	22	.453	.635	05	.351	.141	19	.437	.655
06	.302	.672	21	.318	.088	11	.316	.653	08	.497	.396	21	.472	.824	17	.370	.811	24	.466	.773
01	.409	.406	11	.376	.936	13	.324	.585	25	.503	.893	05	.488	.118	09	.388	.484	14	.531	.014
13	.507	.693	14	.430	.814	12	.351	.275	15	.594	.603	01	.525	.222	04	.410	.073	09	.562	.678
02	.575	.654	27	.438	.676	20	.371	.535	27	.620	.894	12	.561	.980	25	.471	.530	06	.601	.675
18	.591	.318	08	.467	.205	08	.409	.495	21	.629	.841	08	.652	.508	13	.486	.779	10	.612	.859
20	.610	.821	09	.474	.138	16	.445	.740	17	.691	.583	18	.668	.271	15	.515	.867	26	.673	.112
12	.631	.597	10	.492	.474	03	.494	.929	09	.708	.689	30	.736	.634	23	.567	.798	23	.738	.770
27	.651	.281	13	.499	.892	27	.543	.387	07	.709	.012	02	.763	.253	11	.618	.502	21	.753	.614
04	.661	.953	19	.511	.520	17	.625	.171	11	.714	.049	23	.804	.140	28	.636	.148	30	.758	.851
22	.692	.089	23	.591	.770	02	.699	.073	23	.720	.695	25	.828	.425	27	.650	.741	27	.765	.563
05	.779	.346	20	.604	.730	19	.702	.934	03	.748	.413	10	.843	.627	16	.711	.508	07	.780	.534
09	.787	.173	24	.654	.330	22	.816	.802	20	.781	.603	16	.858	.849	19	.778	.812	04	.818	.187
10	.818	.837	12	.728	.523	04	.838	.166	26	.830	.384	04	.903	.327	07	.804	.675	17	.837	.353
14	.895	.631	16	.753	.344	15	.904	.116	04	.843	.002	09	.912	.382	08	.806	.952	05	.854	.818
26	.912	.376	01	.806	.134	28	.969	.742	12	.884	.582	27	.935	.162	18	.841	.414	01	.867	.133
28	.920	.163	22	.878	.884	09	.974	.046	29	.926	.700	20	.970	.582	12	.918	.114	08	.915	.538
03	.945	.140	25	.939	.162	05	.977	.494	16	.951	.601	19	.975	.327	03	.992	.399	25	.975	.584

Col. No. 8			Col. No. 9			Col. No. 10			Col. No. 11			Col. No. 12			Col. No. 13			Col. No. 14		
A	B	C	A	B	C	A	B	C	A	B	C	A	B	C	A	B	C	A	B	C
09	.042	.071	14	.061	.935	26	.038	.023	27	.074	.779	16	.073	.987	03	.033	.091	26	.035	.175
17	.141	.411	02	.065	.097	30	.066	.371	06	.084	.396	23	.078	.056	07	.047	.391	17	.089	.363
02	.143	.221	03	.094	.228	27	.073	.876	24	.098	.524	17	.096	.076	28	.064	.113	10	.149	.681
05	.162	.899	16	.122	.945	09	.095	.568	10	.133	.919	04	.153	.163	12	.066	.360	28	.238	.075
03	.285	.016	18	.158	.430	05	.180	.741	15	.187	.079	10	.254	.834	26	.076	.552	13	.244	.767
28	.291	.034	25	.193	.469	12	.200	.851	17	.227	.767	06	.284	.628	30	.087	.101	24	.262	.366
08	.369	.557	24	.224	.572	13	.259	.327	20	.236	.571	12	.305	.616	02	.127	.187	08	.264	.651
01	.436	.386	10	.225	.223	21	.264	.681	01	.245	.988	25	.319	.901	06	.144	.068	18	.285	.311
20	.450	.289	09	.233	.838	17	.283	.645	04	.317	.291	01	.320	.212	25	.202	.674	02	.340	.131
18	.455	.789	20	.290	.120	23	.363	.063	29	.350	.911	08	.416	.372	01	.247	.025	29	.353	.478
23	.488	.715	01	.297	.242	20	.364	.366	26	.380	.104	13	.432	.556	23	.253	.323	06	.359	.270
14	.496	.276	11	.337	.760	16	.395	.363	28	.425	.864	02	.489	.827	24	.320	.651	20	.387	.248
15	.503	.342	19	.389	.064	02	.423	.540	22	.487	.526	29	.503	.787	10	.328	.365	14	.392	.694
04	.515	.693	13	.411	.474	08	.432	.736	05	.552	.511	15	.518	.717	27	.338	.412	03	.408	.077
16	.532	.112	20	.447	.893	10	.476	.468	14	.564	.357	28	.524	.998	13	.356	.991	27	.440	.280
22	.557	.357	22	.478	.321	03	.508	.774	11	.572	.306	03	.542	.352	16	.401	.792	22	.461	.830
11	.559	.620	29	.481	.993	01	.601	.417	21	.594	.197	19	.585	.462	17	.423	.117	16	.527	.003
12	.650	.216	27	.562	.403	22	.687	.917	09	.607	.524	05	.695	.111	21	.481	.838	30	.531	.486
21	.672	.320	04	.566	.179	29	.697	.862	19	.650	.572	07	.733	.838	08	.560	.401	25	.678	.360
13	.709	.273	08	.603	.758	11	.701	.605	18	.664	.101	11	.744	.948	19	.564	.190	21	.725	.014
07	.745	.687	15	.632	.927	07	.728	.498	25	.674	.428	18	.793	.748	05	.571	.054	05	.797	.595
30	.780	.285	06	.707	.107	14	.745	.679	02	.697	.674	27	.802	.967	18	.587	.584	15	.801	.927
19	.845	.097	28	.737	.161	24	.819	.444	03	.767	.928	21	.826	.487	15	.604	.145	12	.836	.294
26	.846	.366	17	.846	.130	15	.840	.823	16	.809	.529	24	.835	.832	11	.641	.298	04	.854	.982
29	.861	.307	07	.874	.491	25	.863	.568	30	.838	.294	26	.855	.142	22	.672	.156	11	.884	.928
25	.906	.874	05	.880	.828	06	.878	.215	13	.845	.470	14	.861	.462	20	.674	.887	19	.886	.832
24	.919	.300	23	.931	.659	18	.930	.601	08	.855	.524	20	.874	.625	14	.752	.881	07	.929	.937
10	.952	.555	26	.960	.365	04	.954	.827	07	.867	.718	30	.929	.056	09	.774	.560	09	.932	.206
06	.961	.504	21	.978	.194	28	.963	.004	12	.881	.722	09	.935	.582	29	.921	.752	01	.970	.692
27	.969	.811	12	.982	.183	19	.988	.020	23	.937	.872	22	.947	.797	04	.959	.099	23	.973	.082

Table D-1 (Continued). Random Numbers for General Sampling Procedure

Col. No. 15			Col. No. 16			Col. No. 17			Col. No. 18			Col. No. 19			Col. No. 20			Col. No. 21		
A	B	C	A	B	C	A	B	C	A	B	C	A	B	C	A	B	C	A	B	C
15	.023	.979	19	.062	.588	13	.045	.004	25	.027	.290	12	.052	.075	20	.030	.881	01	.010	.946
11	.118	.465	25	.080	.218	18	.086	.878	06	.057	.571	30	.075	.493	12	.034	.291	10	.014	.939
07	.134	.172	09	.131	.295	26	.126	.990	26	.059	.026	28	.120	.341	22	.043	.893	09	.032	.346
01	.139	.230	18	.136	.381	12	.128	.661	07	.105	.176	27	.145	.689	28	.143	.073	06	.093	.180
16	.145	.122	05	.147	.864	30	.146	.337	18	.107	.358	02	.209	.957	03	.150	.937	15	.151	.012
20	.165	.520	12	.158	.365	05	.169	.470	22	.128	.827	26	.272	.818	04	.154	.867	16	.185	.455
06	.185	.481	28	.214	.184	21	.244	.433	23	.156	.440	22	.299	.317	19	.158	.359	07	.227	.277
09	.211	.316	14	.215	.757	23	.270	.849	15	.171	.157	18	.306	.475	29	.304	.615	02	.304	.400
14	.248	.348	13	.224	.846	25	.274	.407	08	.220	.097	20	.311	.653	06	.369	.633	30	.316	.074
25	.249	.890	15	.227	.809	10	.290	.925	20	.252	.066	15	.348	.156	18	.390	.536	18	.328	.799
13	.252	.577	11	.280	.898	01	.323	.490	04	.268	.576	16	.381	.710	17	.403	.392	20	.352	.288
30	.273	.088	01	.331	.925	24	.352	.291	14	.275	.302	01	.411	.607	23	.404	.182	26	.371	.216
18	.277	.689	10	.399	.992	15	.361	.155	11	.297	.589	13	.417	.715	01	.415	.457	19	.448	.754
22	.372	.958	30	.417	.787	29	.374	.882	01	.358	.305	21	.472	.484	07	.437	.696	13	.487	.598
10	.461	.075	08	.439	.921	08	.432	.139	09	.412	.089	04	.478	.885	24	.446	.546	12	.546	.640
28	.519	.536	20	.472	.484	04	.467	.266	16	.429	.834	25	.479	.080	26	.485	.768	24	.550	.038
17	.520	.090	24	.498	.712	22	.508	.880	10	.491	.203	11	.566	.104	15	.511	.313	03	.604	.780
03	.523	.519	04	.516	.396	27	.632	.191	28	.542	.306	10	.576	.659	10	.517	.290	22	.621	.930
26	.573	.502	03	.548	.688	16	.661	.836	12	.563	.091	29	.665	.397	30	.556	.853	21	.629	.154
19	.634	.206	23	.597	.508	19	.675	.629	02	.593	.321	19	.739	.298	25	.561	.837	11	.634	.908
24	.635	.810	21	.681	.114	14	.680	.890	30	.692	.198	14	.749	.759	09	.574	.599	05	.696	.459
21	.679	.841	02	.739	.298	28	.714	.508	19	.705	.445	08	.756	.919	13	.613	.762	23	.710	.078
27	.712	.366	29	.792	.038	06	.719	.441	24	.709	.717	07	.798	.183	11	.698	.783	29	.726	.585
05	.780	.497	22	.829	.324	09	.735	.040	13	.820	.739	23	.834	.647	14	.715	.179	17	.749	.916
23	.861	.106	17	.834	.647	17	.741	.906	05	.848	.866	06	.837	.978	16	.770	.128	04	.802	.186
12	.865	.377	16	.909	.608	11	.747	.205	27	.867	.633	03	.849	.964	08	.815	.385	14	.835	.319
29	.882	.635	06	.914	.420	20	.850	.047	03	.883	.333	24	.851	.109	05	.872	.490	08	.870	.546
08	.902	.020	27	.958	.856	02	.859	.356	17	.900	.443	05	.859	.935	21	.885	.999	28	.871	.539
04	.951	.482	26	.981	.976	07	.870	.612	21	.914	.483	17	.863	.220	02	.958	.177	25	.971	.369
02	.977	.172	07	.983	.624	03	.916	.463	29	.950	.753	09	.863	.147	27	.961	.980	27	.984	.252

Col. No. 22			Col. No. 23			Col. No. 24			Col. No. 25			Col. No. 26			Col. No. 27			Col. No. 28		
A	B	C	A	B	C	A	B	C	A	B	C	A	B	C	A	B	C	A	B	C
12	.051	.032	26	.051	.187	08	.015	.521	02	.039	.005	16	.026	.102	21	.050	.952	29	.042	.039
11	.068	.980	03	.053	.256	16	.068	.994	16	.061	.599	01	.033	.886	17	.085	.403	07	.105	.293
17	.089	.309	29	.100	.159	11	.118	.400	26	.068	.054	04	.088	.686	10	.141	.624	25	.115	.420
01	.091	.371	13	.102	.465	21	.124	.565	11	.073	.812	22	.090	.602	05	.154	.157	09	.126	.612
10	.100	.709	24	.110	.316	18	.153	.158	07	.123	.649	13	.114	.614	06	.164	.841	10	.205	.144
30	.121	.744	18	.114	.300	17	.190	.159	05	.126	.658	20	.136	.576	07	.197	.013	03	.210	.054
02	.166	.056	11	.123	.208	26	.192	.676	14	.161	.189	05	.138	.228	16	.215	.363	23	.234	.533
23	.179	.529	09	.138	.182	01	.237	.030	18	.166	.040	10	.216	.565	08	.222	.520	13	.266	.799
21	.187	.051	06	.194	.115	12	.283	.077	28	.248	.171	02	.233	.610	13	.269	.477	20	.305	.603
22	.205	.543	22	.234	.480	03	.286	.318	06	.255	.117	07	.278	.357	02	.288	.012	05	.372	.223
28	.230	.688	20	.274	.107	10	.317	.734	15	.261	.928	30	.405	.273	25	.333	.633	26	.385	.111
19	.243	.001	21	.331	.292	05	.337	.844	10	.301	.811	06	.421	.807	28	.348	.710	30	.422	.315
27	.267	.990	08	.346	.085	25	.441	.336	24	.363	.025	12	.426	.583	20	.362	.961	17	.453	.783
15	.283	.440	27	.382	.979	27	.469	.786	22	.378	.792	08	.471	.708	14	.511	.989	02	.460	.916
16	.352	.089	07	.387	.865	24	.473	.237	27	.379	.959	18	.473	.738	26	.540	.903	27	.461	.841
03	.377	.648	28	.411	.776	20	.475	.761	19	.420	.557	19	.510	.207	27	.587	.643	14	.483	.095
06	.397	.769	16	.444	.999	06	.557	.001	21	.467	.943	03	.512	.329	12	.603	.745	12	.507	.375
09	.409	.428	04	.515	.993	07	.610	.238	17	.494	.225	15	.640	.329	29	.619	.895	28	.509	.748
14	.465	.406	17	.518	.827	09	.617	.041	09	.620	.081	09	.665	.354	23	.623	.333	21	.583	.804
13	.499	.651	05	.539	.620	13	.641	.648	30	.623	.106	14	.680	.884	22	.624	.076	22	.587	.993
04	.539	.972	02	.623	.271	22	.664	.291	03	.625	.777	26	.703	.622	18	.670	.904	16	.689	.339
18	.560	.747	30	.637	.374	04	.668	.856	08	.651	.790	29	.739	.394	11	.711	.253	06	.727	.298
26	.575	.892	14	.714	.364	19	.717	.232	12	.715	.599	25	.759	.386	01	.790	.392	04	.731	.814
29	.756	.712	15	.730	.107	02	.776	.504	23	.782	.093	24	.803	.602	04	.813	.611	08	.807	.983
20	.760	.920	19	.771	.552	29	.777	.548	20	.810	.371	27	.842	.491	19	.843	.732	15	.833	.757
05	.847	.925	23	.780	.662	14	.823	.223	01	.841	.726	21	.870	.435	03	.844	.511	19	.896	.464
25	.872	.891	10	.924	.888	23	.848	.264	29	.862	.009	28	.906	.367	30	.858	.299	18	.916	.384
24	.874	.135	12	.929	.204	30	.892	.817	25	.891	.873	23	.948	.367	09	.929	.199	01	.948	.610
08	.911	.215	01	.937	.714	28	.943	.190	04	.917	.264	11	.956	.142	24	.931	.263	11	.976	.799
07	.946	.065	25	.974	.398	15	.975	.962	13	.958	.990	17	.993	.989	15	.939	.947	24	.978	.633

Appendix E. Modified Hveem Method for Emulsified Asphalt-Aggregate Cold Mixture Design

E.01 SCOPE.—This method covers the selection, proportioning and testing of aggregates, additives and emulsified asphalt for dense-graded mixes for pavement construction. It contains California Department of Transportation test methods or modifications of these methods as well as procedures developed within the Asphalt Institute. Criteria to determine the suitability of emulsified asphalt mixes are presented. Procedures for resilient modulus determination are, however, applicable to mixes using emulsified asphalt or paving grade asphalts.

For information on mix-design methods when using paving asphalt (asphalt cement), refer to the Asphalt Institute publication, *Mix Design Methods for Asphalt Concrete and other Hot-Mix Types,* Manual Series No. 2 (MS-2).

E.02 OUTLINE OF METHOD.—

a. *General*
For convenience, the design method is divided into these parts:
(1) Selection of aggregate and emulsified asphalt
(2) Trial emulsified asphalt content
(3) Mixing test (determination of optimum fluids content at mixing)
(4) Determination of optimum fluids content for compaction
(5) Strength testing
(6) Moisture exposure and stability and cohesion testing
(7) Determination of optimum emulsified asphalt content

See Figure E-1 for the testing schedule for this mix design method.

b. *Selection of Aggregate and Emulsified Asphalt*
Aggregates used for emulsified asphalt paving mixtures and guidelines for the selection of emulsified asphalt type are both discussed in Chapter II of this manual.

c. *Trial Emulsified Asphalt Content*
The Centrifuge Kerosene Equivalent Test (C.K.E.) is used for estimating the emulsified asphalt contents for trial mixes of aggregates. Ranges of emulsified asphalt content for trial mixes are shown in Table E-1.

d. *Mixing Test, Determination of Optimum Fluids Content at Mixing*
Either a spoon and bowl or mechanical mix is made to determine the coating and workability of the trial mixtures. The amount of mix water is varied to optimize these properties unless job conditions obviously prevent such optimization. Additives, if used, are premixed with the aggregate prior to conducting the mixing test.

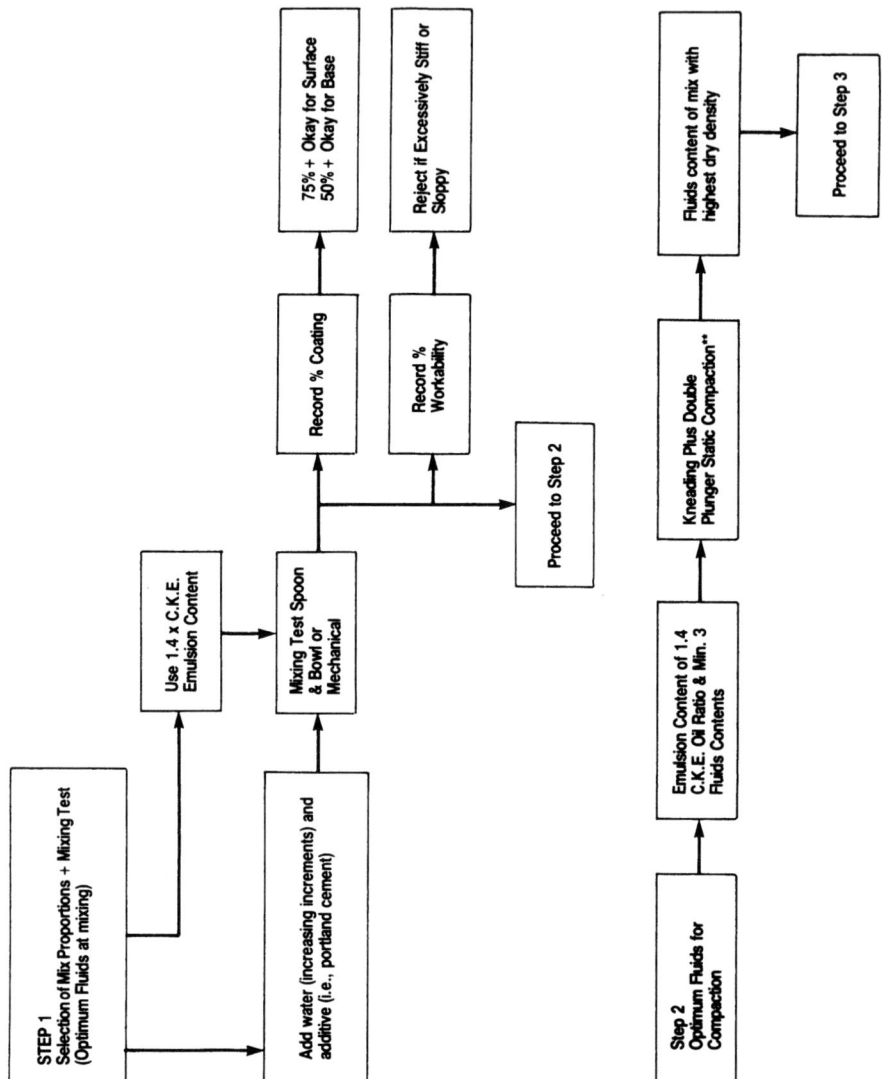

Figure E-1. Testing schedule for dense-graded emulsified asphalt mixes.

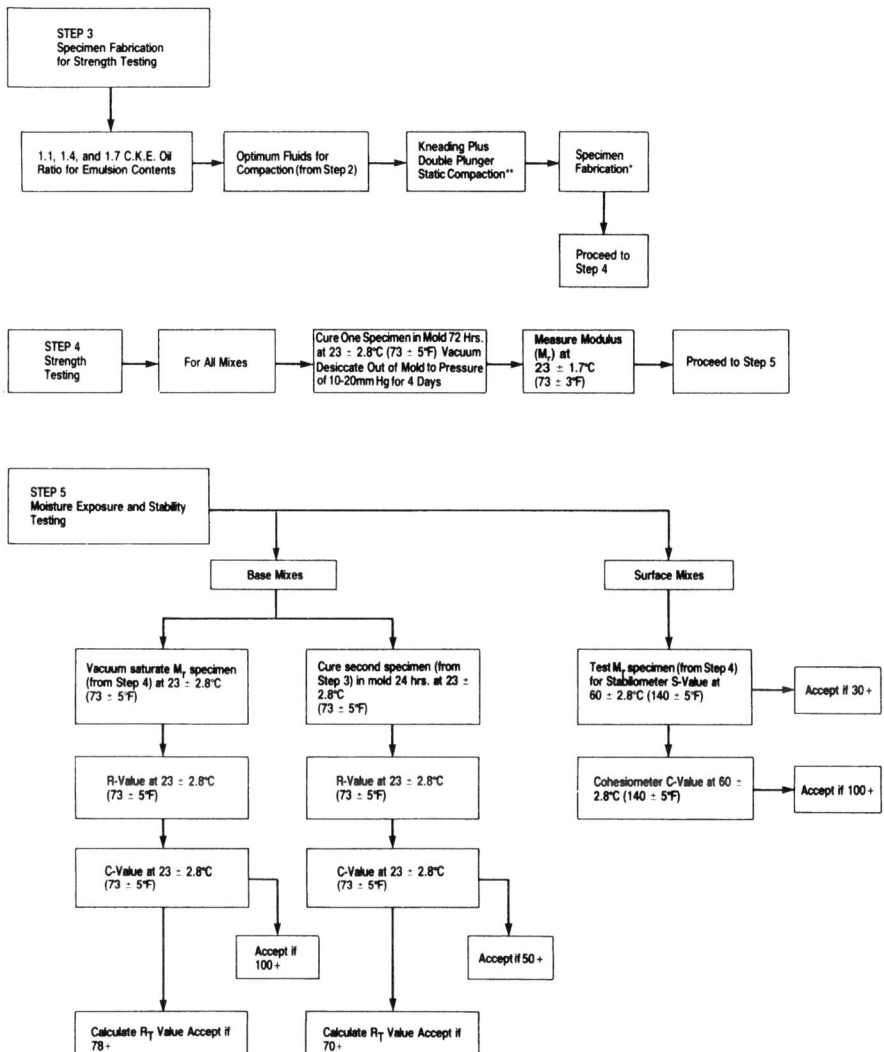

Figure E-1 (Cont.). Testing schedule for dense-graded emulsified asphalt mixes.

e. Optimum Fluids Content for Compaction

Determination of the optimum fluids content (mixing water plus emulsified asphalt) for compaction and test specimen fabrication are achieved by a light kneading compaction followed by a double plunger static load.

f. Strength Testing

The strength of emulsified asphalt mixes is measured by running a final modulus at a temperature of $23 \pm 1.7°$ C ($73 \pm 3°$ F) after a total of three days mold cure plus four days vacuum desiccation.** This data is used in conjunction with certain project variables (traffic, regional temperature and curing conditions) and other mix properties (volume percent of asphalt residue and air voids) in determining the pavement thickness requirements.

g. Moisture Exposure and Stability Testing

Base mixes have their strength evaluated before and after vacuum saturation. Base mixes are tested at $23 \pm 2.8°$ C ($73 \pm 5°$ F) for Resistance R-Value and Cohesiometer C-Value. Surface mixes are tested at $60 \pm 2.8°$ C ($140 \pm 5°$ F) for Stabilometer S-Value and Cohesiometer C-Value.

h. Determination of Optimum Emulsion Content

Table E-5 gives design criteria for the two types of emulsified asphalt dense-graded mixes.

Table E-1. Selection of Emulsified Asphalt Content

Aggregate Type	Approximate Emulsified Asphalt Content, Percent by Weight of Aggregate*
Processed Dense Graded	5.0 - 10.0
Sands Silty Sands Semi-Processed Crusher Pit or Bank Run	4.5-8.0

*With porous aggregates the emulsified asphalt content should be increased by a factor of approximately 1.2. Porous aggregates are those which absorb more than 2 percent water by dry weight when tested by ASTM Method C 127.

**An alternate procedure, that shortens the laboratory curing time, is to cure for one day in mold at room temperature followed by one day out of mold in oven at $37.8°$ C ($100°$ F).

E.03 AGGREGATES FOR EMULSIFIED ASPHALT MIXES.—

a. *General*

The types of materials that are suitable for emulsified asphalt treatment include sand, blast furnace slag, coral, volcanic cinder, gravel, ore tailings, crushed ledge stone or rock, reclaimed aggregate or other inert material.

b. *Selection*

Aggregates meeting the requirements described earlier are among those suitable for emulsified asphalt mixes. All of these aggregates are acceptable for bases and also for temporary surfaces for at least light traffic. (For mix design purposes, temporary surfaces are treated as base mixes.) However, for permanent surfaces, the processed dense-graded or open-graded aggregates plus a surface treatment will be required.

E.04 ASPHALTS.—Two types of emulsified asphalt are used for mixing. These are designated as slow setting (SS) and medium setting (MS).

E.05 DETERMINING MIX PROPORTIONS.—

a. *General*

The amount of emulsified asphalt is estimated for trial mixes of dense-graded aggregates using the Centrifuge Kerosene Equivalent (C.K.E.) test.

b. *Centrifuge Kerosene Equivalent Test*
(1) General

The first step in this method of mix design is to determine the approximate asphalt content by the Centrifuge Kerosene Equivalent method.* With a calculated surface area and the factors obtained by the C.K.E. method for a particular aggregate or blend of aggregates, the approximate asphalt content is determined by using a series of charts. These charts are presented herein, accompanied by typical examples to demonstrate their application.

(2) Equipment

The equipment and materials required for determining the approximate asphalt content are:
(a) *Sample Splitter*, small, for obtaining representative samples of fine aggregate.

---

*The development of this method of determining optimum asphalt content is outlined in "Establishing the Oil Content for Dense-Graded Bituminous Mixtures" by F. N. Hveem, *California Highways and Public Works*, July-August, 1942.

(b) *Pans*, 114 mm (4 1/2 in.) diameter × 25 mm (1 in.) deep.
(c) *Kerosene*, 4 litres (1 gal.).
(d) *Oil*, SAE No. 10, lubricating, 4 litres (1 gal.).
(e) *Beakers*, 1500 ml.
(f) *Metal Funnels*, 89 mm (3 1/2 in.) top diameter, 114 mm (4 1/2 in.) height, 13 mm (1/2 in.) orifice with piece of 2.00 mm (No. 10) sieve soldered to bottom of opening.
(g) *Timer.*
(h) *Centrifuge*, hand-operated, complete with cups, capable of producing 400 times gravity (a power-driven centrifuge is available from Soiltest, Inc., 2205 Lee Street, Evanston, Illinois 60602, Catalog No. AP-275 or equivalent).
(i) *Filter Papers*, 55 mm diameter (No. 611, Eaton-Dikeman Co., Mt. Holly Springs, Pennsylvania), or equivalent.

(3) Surface Area

The gradation of the aggregate or blend of aggregates employed in the mix is used to calculate the surface area of the aggregates. This calculation consists of multiplying the total percent passing each sieve by a "surface-area factor" as set forth in Table E-2. Add the products thus obtained and the total will represent the equivalent surface area of the sample in terms of m^2/kg (ft^2/lb). It is important to note that all surface-area factors must be used in the calculation. Also, if a different series of sieves is used, different surface-area factors are necessary.

Table E-2. Surface Area Factors

Total Percent Passing Sieve No.	Maximum Size	4.75 mm (No. 4)	2.36 mm (No. 8)	1.18 mm (No. 16)	600 μm (No. 30)	300 μm (No. 50)	150 μm (No. 100)	75 μm (No. 200)
Surface Area Factor,* m^2/kg (ft^2/lb.)	.41 (2)	.41 (2)	.82 (4)	1.64 (8)	2.87 (14)	6.14 (30)	12.29 (60)	32.77 (160)

*Surface area factors shown are applicable only when all the above-listed sieves are used in the sieve analysis.

This tabulation demonstrates the calculation of surface area by this method:

Sieve Size	Percent Passing	x	S.A. Factor	=	Surface Area
19.0 mm (3/4 in.)	100 ⎫		.41 (2)		.41 (2)
9.5 mm (3/8 in.)	90 ⎭				
4.75 mm (No. 4)	75		.41 (2)		.31 (1.5)
2.36 mm (No. 8)	60		.82 (4)		.49 (2.4)
1.18 mm (No. 16)	45		1.64 (8)		.74 (3.6)
600 μm (No. 30)	35		2.87 (14)		1.00 (4.9)
300 μm (No. 50)	25		6.14 (30)		1.54 (7.5)
150 μm (No. 100)	18		12.29 (60)		2.21 (10.8)
75 μm (No. 200)	10		32.77 (160)		3.28 (16.0)

Surface Area 9.98 m^2/kg
(48.7 ft^2/lb)

(4) C.K.E. Procedure
   (a) Place exactly 100g of dry aggregate (representative of the passing 4.75 mm [No. 4] material being used) in the tared centrifuge cup assembly fitted with a screen and a disk of filter paper.
   (b) Place bottom of centrifuge cup in kerosene until the aggregate becomes saturated.
   (c) Centrifuge the saturated sample for 2 minutes at a force of 400 times gravity. (For the suggested centrifuge this force can be developed by turning the handle approximately 45 revolutions per minute.)
   (d) Weigh sample after centrifuging and determine the amount of kerosene retained as a percent of the dry aggregate weight; this value is called the Centrifuge Kerosene Equivalent (C.K.E.). (Note: Duplicate samples are always prepared in order to balance the centrifuge and to check results. The average of the two C.K.E. values is used unless there is a large discrepancy, in which case the test is rerun.)
   (e) If the apparent specific gravity of samples is greater than 2.70 or less than 2.60 make a correction to the C.K.E. value using the formula at the bottom of the chart in Figure E-2.

(5) Surface Capacity Test (Oil Soak) for Coarse Aggregate
   (a) Place into a metal funnel exactly 100g of dry aggregate passing the 9.5 mm (3/8 in.) sieve and retained on the 4.75 mm (No. 4) sieve (this fraction is considered to be representative of the coarse aggregate in the mix.)
   (b) Immerse sample and funnel in a beaker containing SAE No. 10 lubricating oil at room temperature for 5 minutes.
   (c) Drain for 2 minutes.

(d) Remove funnel and sample from oil and drain for 15 minutes at a temperature of 60° C (140° F).
(e) Weigh the sample after draining and determine the amount of oil retained as a percent of the dry aggregate weight. (Note: Duplicate samples are prepared to check results. Average value is used unless there is a large discrepancy, in which case the test is rerun.)
(f) If the apparent specific gravity is greater than 2.70 or less than 2.60 make a correction to the percent oil retained using the formula at the bottom of the chart in Figure E-3.

(6) Estimated Optimum Emulsified Asphalt Content
  (a) Using the C.K.E. value obtained and the chart in Figure E-2, determine the value K_f (surface constant for fine material).
  (b) Using the percent oil retained and the chart in Figure E-3, determine the value K_c (surface constant for coarse material).
  (c) Using the values obtained for K_f and K_c and chart in Figure E-4, determine the value K_m (surface constant for fine-coarse aggregate combined). $K_m = K_f +$ correction to K_f. The correction to K_f obtained from Figure E-4 is positive if ($K_c - K_f$) is positive and is negative if ($K_c - K_f$) is negative.
  (d) The next step is to determine the approximate asphalt ratio for the mix based on cutback asphalts of RC-250, MC-250 and SC-250 grades. With values obtained for K_m, surface area and average specific gravity use the Case 2 procedures of the chart in Figure E-5 to determine the oil ratio.

(7) Example
To demonstrate the use of the charts in Figures E-2 through E-5, assume these conditions apply to a paving mix using emulsified asphalt:

Apparent Specific Gravity, coarse = 2.45
Apparent Specific Gravity, fine = 2.64
Percent Passing 4.75 mm (No. 4) = 45

$$\text{Avg. Sp. Gr.} = \frac{100}{\frac{55}{2.45} + \frac{45}{2.64}} = 2.53$$

Surface Area of Aggregate Grading = 6.6 m²/kg (32.4 ft²/lb)
C.K.E. = 5.6
Percent Oil Retained, coarse = 1.9
(corrected for specific gravity, this value is 1.7 percent. See Figure E-3)

From Figure E-2 determine K_f as 1.25.
From Figure E-3 determine K_c as 0.8.
From Figure E-4 determine K_m as 1.15.
From Figure E-5 determine the oil ratio for liquid asphalt as 5.2 percent.

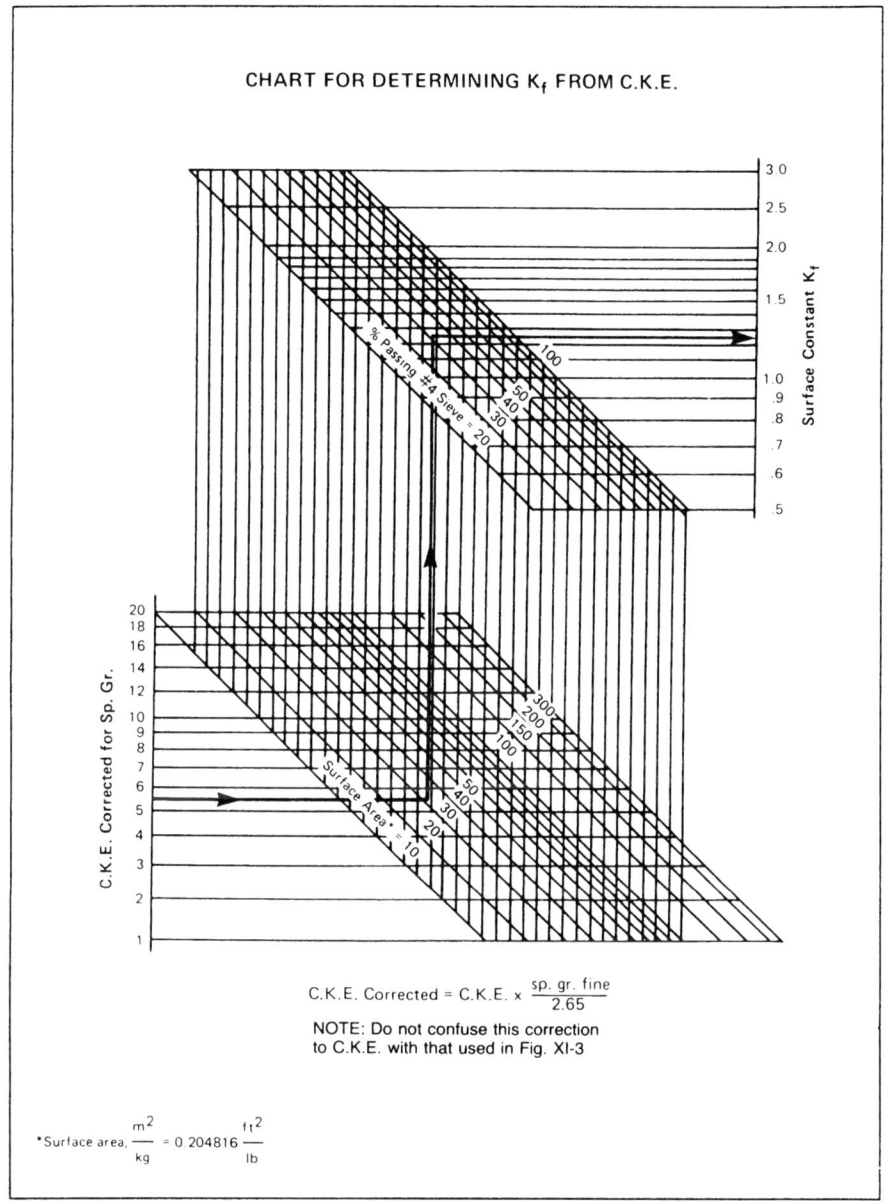

Figure E-2. Chart for determining surface constant for fine material, K_f, from C.K.E., Hveem method of design.
Chart courtesy of California Department of Transportation.

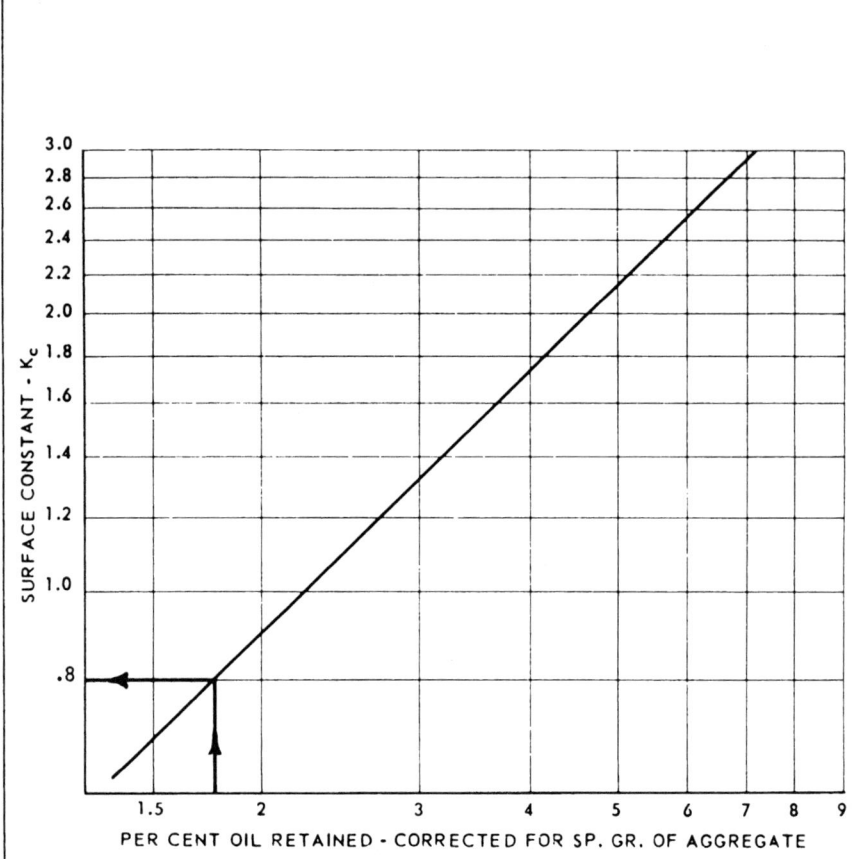

Figure E-3. Chart for determining surface constant for coarse material, K_c, from coarse aggregate absorption, Hveem method of design.
Chart courtesy of California Department of Transportation.

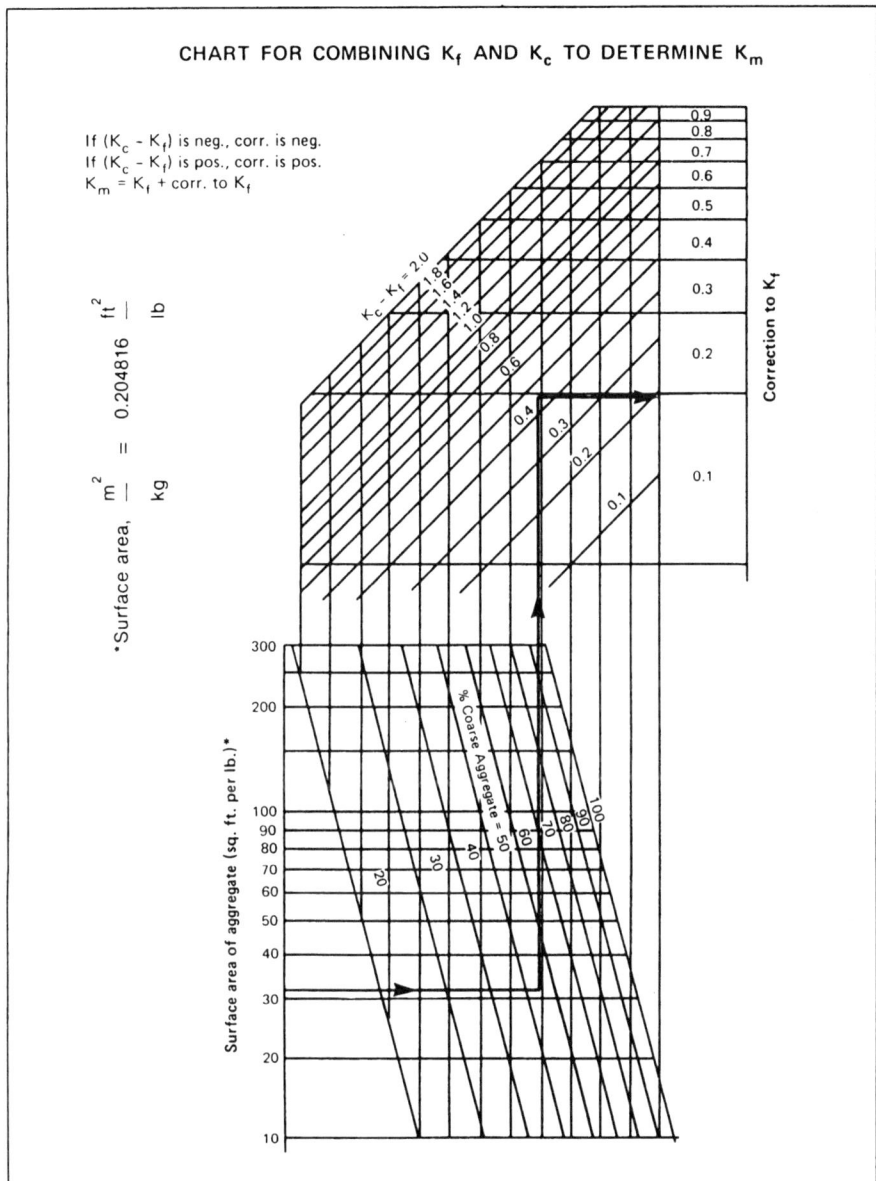

Figure E-4. Chart for combining K_f and K_c to determine surface constant for combined aggregate, K_m, Hveem method of design.
Chart courtesy of California Department of Transportation.

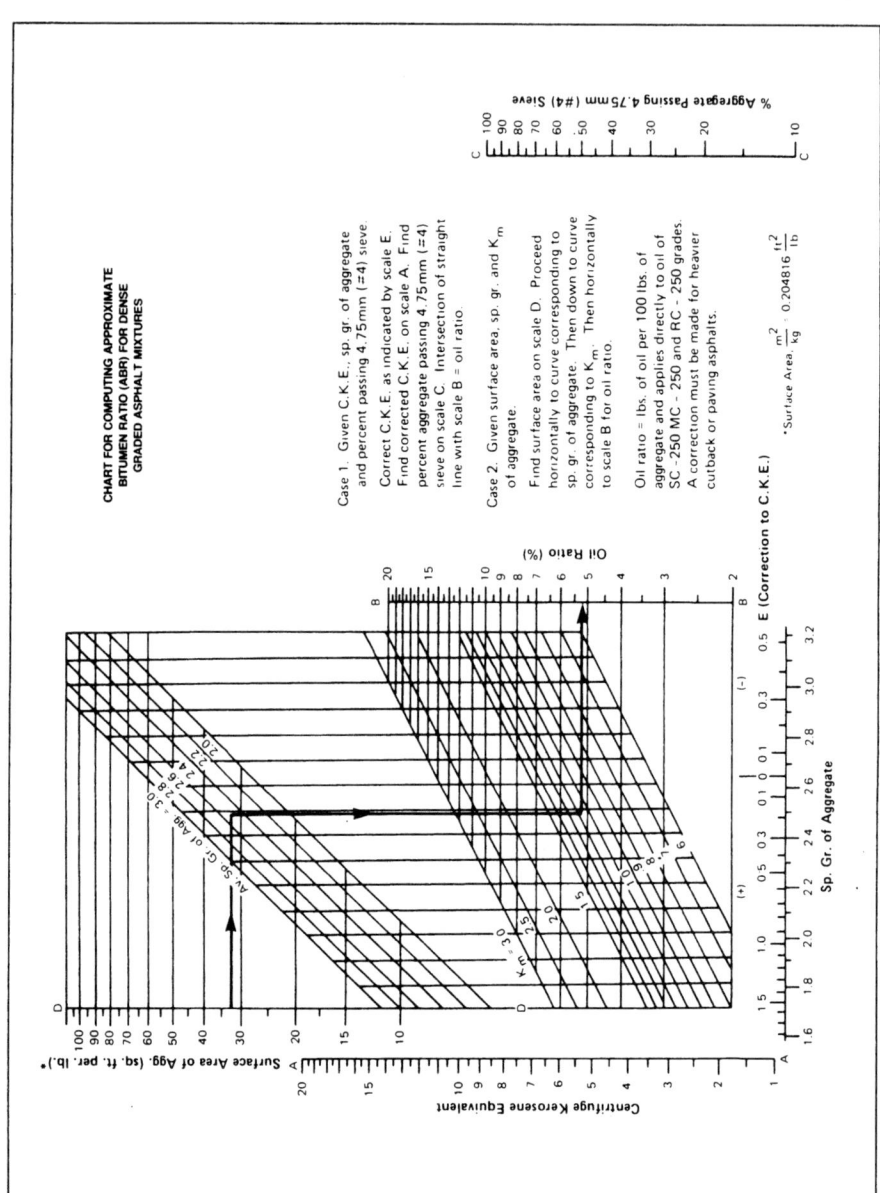

Figure E-5. Chart for computing oil ratio for dense-graded asphalt mixtures, Hveem method of design.
Chart courtesy of California Department of Transportation.

c. *Emulsified Asphalt Content for Trial Mixes*

The trial emulsified asphalt content for dense-graded mixes is equal to 1.4 × C.K.E. oil ratio, and is adjusted to a 60 percent residue as follows:

$$\frac{\text{Correct Emulsified}}{\text{Asphalt Content}} = \frac{(1.4 \times \text{C.K.E. Oil Ratio}) \times 60}{\text{Emulsion Residue \%}}$$

d. *Mixing Test, Determination of Optimum Fluids Content at Mixing*

(1) General

This test measures the ability of the emulsified asphalt to uniformly disperse throughout the mix. It also allows the laboratory technician to judge the mix workability. A number of variables have been found to influence *asphalt dispersion* and these are listed in Table E-3.

Table E-3. Variables Affecting Asphalt Dispersion

Variable	Factors Influencing	To Improve Asphalt Dispersion
Aggregate	Surface area (fines)	Decrease
	Porosity	Decrease
	Roughness	Decrease
Emulsified Asphalt	Amount	Increase
	Type	Use anionics with calcareous aggregates (limestone).
		Use cationics with siliceous aggregates.
	Oil Distillate	Increase with MS-types.
Mixing Water	Amount	Increase with SS-types.
		Decrease with most MS-types.
		Decrease to reduce asphalt runoff.
Mixing Operations	Temperature	Decrease with SS-types to prevent coalescence during the mixing cycle.
		Increase with MS-types.
	Mix Cycle	Optimize. Insufficient mixing may give poor coating.
		Excessive mixing may induce stripping.

84 Appendix E

Figure E-6. Apparatus for Hveem C.K.E. tests.

(2) Equipment
   (a) *Balance*, 5,000 g minimum capacity and accurate to within ± 0.5 g.
   (b) *Mixing Equipment*, preferably mechanized and capable of producing intimate mixtures of the job aggregate, water and asphalt. Hand mixing, if used, must be sufficiently thorough to uniformly disperse the water and emulsified asphalt throughout the aggregate.
   (c) *Hot Plate* or 110 ± 5° C (230 ± 9° F) oven.
   (d) Supply of *round bottom mixing bowls* (approximately 5 litre (5 qt.) capacity).
   (e) Supply of 250 mm (10 in.) *metal kitchen mixing spoons*.
   (f) *One-hundred millilitre glass graduate*.

(3) Procedure
   (a) Obtain representative samples of each emulsified asphalt to be considered for the project.
   (b) Obtain representative samples of the job aggregate or aggregate blend.
   (c) Determine the moisture content on the aggregate according to ASTM D 2216 and record.

(d) Prepare the remainder of the aggregate by drying to constant weight at 60° C (140° F). Separate into sizes using these sieves: 25.0 mm, 19.0 mm, 12.5 mm, 9.5 mm, and 4.75 mm (1 in., 3/4 in., 1/2 in., 3/8 in. and No. 4).

(e) Weigh out a sufficient number of batches of the job aggregate for mixing tests. The batch weight shall be based on the nominal maximum size particle in the aggregate:

Nominal Maximum Particle Size	Batch Weight
25.0 mm (1 in.)	2,000 grams minimum
19.0 mm (3/4 in.)	1,200 grams minimum
12.5 mm (1/2 in.)	750 grams minimum
4.75 mm (No. 4)	500 grams minimum

Note: These batches should be prepared by reblending exact fractions of plus 4.75 mm (No. 4) material with minus 4.75 mm (No. 4) material to match the grading analysis of the whole sample.

(f) Put one batch of aggregate in the mixing bowl and incorporate the additive (i.e., cement) if specified.

(g) Add and incorporate the minimum amount of mixing water required to achieve coating. Normally this is just enough to darken the aggregate.

Note: In areas where the addition or removal of water is uneconomical, mixes should be made at the in-situ moisture content.

(h) The emulsified asphalt, in an amount as determined by the "correct emulsified asphalt content" of Art. E.05 (c), is added to the damp aggregate and mixed. The mix cycle should simulate field mixing operations (generally a 1-minute cycle* with a laboratory mechanical mixer or a 2-minute spoon bowl mix is sufficient). Judge the suitability of the finished mix by the uniformity of the color (best judged by drying a small portion of the batch on a hot plate). Spottiness denotes an unsatisfactory mix (usually due to insufficient water or improper mixing properties of the emulsion). Mixes which strip or stiffen excessively on mixing are also considered unsatisfactory. If unsatisfactory, rerun a new batch with an additional increment of water and observe for suitability as before. Repeat until a satisfactory-appearing mix is obtained. Mixes which become excessively soupy with additional water and segregate on standing are considered unsatisfactory.

(i) Selection of the emulsified asphalt for the project shall be based upon these considerations:

(1) *Coating.* As close as possible to 100 percent coating is preferred. Mixes will be considered suitable if they have a minimum of 75

---

*Mixing time may be shortened to 30 seconds if segregation in the mixture is noticed.

percent coating if used as a surface and 50 percent if used as a base.
  (2) *Workability.* The mix should be workable. Mixes which are excessively stiff or soupy should be rejected.
  (3) *Job Conditions.* The availability of water at the construction site, mixing process and anticipated rate of the emulsion mixture cure will also influence the selection of the type and grade of emulsified asphalt.
(j) The total fluids content at mixing of the satisfactory mix is computed by adding percentages of the asphalt emulsion, added mixing water and natural water content of the aggregate. It is expressed as weight percent of dry aggregate. This optimum fluids content at mixing is used for establishing weight proportions of subsequent batches.

E.06 OPTIMUM FLUIDS CONTENT FOR COMPACTION.—

a. *General*

The optimum fluids content for compaction is determined as well as the fabrication of all test specimens using a light kneading compaction followed by a double plunger static load.

b. *Equipment*
(1) *Mechanical compactor,* meeting the requirements of ASTM Method D 1561, "Compaction of Test Specimens of Bituminous Mixtures by Means of California Kneading Compactor."
(2) *Compactor accessories*; 101.6 mm (4 in.) diameter × 127 mm (5 in.) high stainless steel molds, and a mold holder.
(3) *Compression testing machine,* 222 kN (50,000 lb.) capacity.
(4) *Two follower rams;* one ram 101.2 mm (3.985 in.) outside diameter × 139.7 mm (5.5 in.) high, and the other ram 101.2 mm (3.985 in.) outside diameter × 38.1 mm (1.5 in.) high.
(5) *Special feeder trough,* 100 mm (4 in.) wide and 405 mm (16 in.) long.
(6) *Metal paddle* to fit feeder trough.
(7) *Bullet-nosed steel rod,* 9.5 mm (3/8 in.) diameter by 405 mm (16 in.) long.
(8) *A device for measuring the height of test specimens* to the nearest 0.25 mm (0.01 in.).
(9) *Metal plate,* 6.4 mm (1/4 in.) thick, approximately 9.5 mm (3/8 in.) wide and 63.5 mm (2 1/2 in.) long.

c. *Optimum Fluids for Compaction*
(1) Prepare a mix at the fluids content as developed by the procedures of Art. E.05.
(2) Weigh at least three batches of approximately 1,200 grams each of this mix for the fluids-density curve specimens. One batch will be compacted

immediately with the remaining batches loose-cured (aerated) prior to compaction so as to produce lower fluids contents (minimum of 3 points required to establish optimum). As a first step in the compaction of all specimens, the mix will be spread uniformly on the feeder trough as shown in Figure E-7. Using paddle, push one-half of the mix into the mold. Rod the mix 20 times in the center of the mass and 20 times around the edge with the bullet-nosed steel rod (Figure E-8). Then, push the remainder of the mix into the mold and repeat the rodding procedure.

(3) Place the mold in the mold holder. Slide the metal insert plate [Art. E.06 b. (9)] under the bottom edge of mold. Note: This is used to give temporary support during the preliminary compaction step.

(4) Place the mold holder containing the mix into the kneading compactor (Figure E-9).

(5) Start the compactor and adjust the tamper foot pressure to 1.7 M Pa (250 psi). Apply approximately 20 tamping blows at 1.7 M Pa (250 psi) pressure to accomplish preliminary consolidation of the mix. The exact number of blows to accomplish the initial compaction shall be determined by observation. The number of blows may vary between 10 and 50, depending upon the type of material. In some instances with sandy or unstable material, it may not be possible to accomplish compaction in the mechanical compactor because of the undue movement of these mixtures under the compactor foot. Discontinue compaction if the compactor foot penetrates more than 6.5 mm (1/4 in.) or if fluids exude from the base of the compaction mold, and proceed to step (6).

Note: Occasionally, the mix may adhere to the compactor foot. When this occurs, stop the compactor and clean the foot. Use heat on the compactor foot only if necessary to prevent sticking.

(6) Remove the mold from the holder and apply a 178 kN (40,000 lb.) static load by the double plunger method, in which a free-fitting plunger is placed below the sample as well as on top. Load at a rate of about 1.3 mm/min (0.05 in./min) and maintain the full load for one minute and release. Reduce the level of the static load if excess fluids exude from the compaction mold and proceed to step (7).

(7) Specimens shall be cured for one day in the mold at room temperature and then extruded. After extrusion, the bulk specific gravities of the specimens are determined by displacement in water (ASTM D 1188 or D 2726).

(8) A plot is made of dry density versus fluids content at compaction. The fluids content resulting in the highest dry density is optimum for compaction.

Figure E-7. Transfer of mix to mold.

Figure E-8. Rodding mix in mold.   **Figure E-9. Mechanical kneading compactor.**

d. *Specimen Fabrication for Strength Testing*
(1) Use approved mix as developed following the procedures of Art. E.05 for strength test specimens. This would include trial mixes at emulsified asphalt contents of 1.1, 1.4 and 1.7 times the C.K.E. Oil Ratio adjusted to a 60 percent residue.
(2) Weigh sufficient mix (approximately 1,150 grams) to fabricate each 101.6 mm (4 in.) diameter specimen, 63.5 mm (2.5 in.) in height, and loose cure (aerate) the mix to the optimum fluids content determined by the procedures of Art. E.06c. Due to differing test procedures, two specimens will be required at each of the trial emulsified asphalt contents selected for bases and temporary wearing surfaces, while only one specimen for each trial emulsified asphalt content is required for permanent surfaces.

E.07 STRENGTH TESTING.—

a. *General*
The rate at which emulsified asphalt mixes cure or develop tensile strength is important. A number of factors including the aggregate gradation, type and amount of emulsion, type and amount of additive, construction and climatic conditions must be assessed by the engineer in determining the rate of tensile strength development. This procedure measures one strength parameter of the mix, the final modulus, using the Resilient Modulus Test. The curing procedure used in defining this strength is given.

b. *Equipment*
Resilient Modulus apparatus and support equipment as manufactured by Retsina Company, 601 Brush Street, Oakland, California 94607.

c. *Procedure*
(1) One of the two specimens compacted at each asphalt content using the procedures of Art E.06 will be cured by placing the mold in a horizontal position for a total of 72 hours at a temperature of $23 \pm 2.8°$ C ($73 \pm 5°$ F).
(2) Remove the specimen(s) from the mold and vacuum desiccate for 4 days (Figure E-11). Adjust the total pressure to 10-20 mm of Hg. (Note: Fill the bottom of the desiccator with Drierite to facilitate removal of water. If the Drierite is spent, indicated by pink color, replace with fresh Drierite and desiccate for an additional day.)
(3) Determine the specimen's final modulus (M_f) at $23 \pm 1.7°$ C ($73 \pm 3°$ F) as outlined below.
  (a) Resilient Modulus Test, M_r.
    The equipment used for determining M_r consists of a repetitive loading device which applies a 0.1 second pulsed load every 3 seconds across the diameter of the test specimen (Figure E-10). The horizontal response to the applied load is measured by a pair of transducers mounted in a yoke that is clamped to the specimen (Figure E-11). The

electrical output from the transducers is amplified and shown on a recording meter designed to hold the deflection long enough for the operator to record the reading.

(b) Calibration

Primary calibration of the instrument is made with a Lansing Instrument Differential Translator. Secondary and load calibration is made with a proving ring. Calibration details are supplied by the instrument manufacturer.

(c) Temperature Control

The M_r of an asphalt treated mix is in part dependent on the stiffness of the asphalt binder. Consequently the M_r is quite dependent on the temperature of the specimen. Equilibrate all specimens and the yoke assembly at least 2 hours at a temperature of $23 \pm 1.7°$ C ($73 \pm 3°$ F) before testing.

(d) M_r Measurements

(1) Place the yoke assembly on the holder (Figure E-12).

(2) Back out the thumb screw so that the transducer levers are clear. Back out the four clamping screws and gently insert the 100 mm (4 in.) diameter sample into the center of the yoke. Place the sample squarely on the centering strip. Gently tighten the four clamping screws, keeping the sample centered and square in the yoke. Use only enough pressure to keep the yoke from falling off the sample (Figure E-13).

(3) Place the assembly in the loading device and align on the centering strip. (Note: Do not lift by the yoke.)

(4) Lift the loading shaft and place the top loading block on the specimen, 3.14 radians (180°) from the bottom centering strip. Allow the shaft to seat against the ball on top of the loading block (Figure E-14).

(5) Zero the recording meter. Set the multiplier knob to 100 and turn on the meter. Adjust the zero control until the meter reads just above zero (Figure E-15).

(6) Tighten one of the transducer advancement screws until an increased meter reading of about 1.0 is obtained. Tighten the advancement screw on the other transducer until an additional increase of 1.0 is obtained on the meter.

(7) Set the pressure regulator to the desired load (Figure E-16). Some instruments use manometers to measure the air pressure—others use pressure transducers. Choice of load depends on the strength of the specimen. Usually 0.33 kN (75 lb.) is used on sound dry specimens having modulus values ranging from 6.89×10^5 to 3.45×10^7 kPa (10^5 to 5×10^6 psi). However, lower pressures may be required to minimize specimen damage. M_r values as low as 3.45×10^4 kPa (5×10^3 psi) can be measured.

(8) Reset the zero knob to just above zero, i.e., until both the high and low pilot lights are out.
(9) Set the mode switch to operate.
(10) Record the deflection in microinches on the meter. If the reading is out of range change the multiplier to a higher or lower value. Reset the zero knob if one of the zero indicator pilot lights is on and make another measurement.
(11) More complete operating details are supplied by the instrument manufacturer.
(12) Rotate the sample 1.57 rad (90°) and repeat measurements. Deflection readings should normally agree within 10 percent. Sometimes a specimen is non-isotropic and a larger difference exists.

Figure E-10. Resilient modulus device.

Figure E-11. Transducers and M_r yoke.

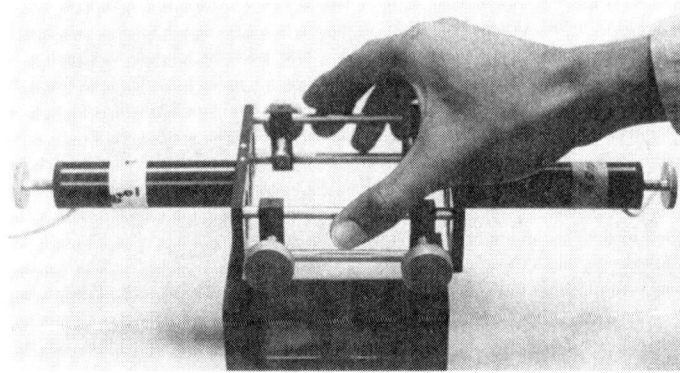

Figure E-12. M_r yoke on holder.

Figure E-13. Tightening M_r clamping screws.

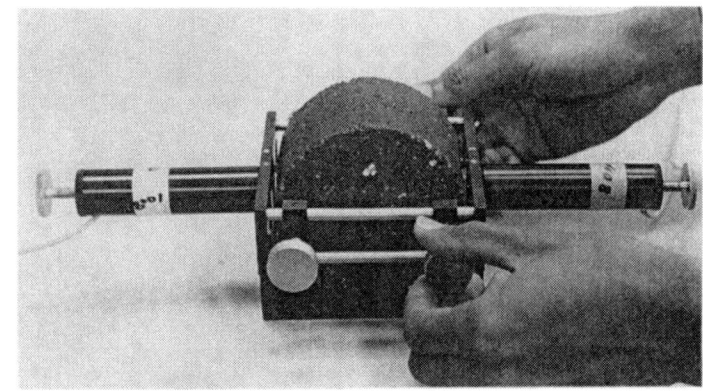

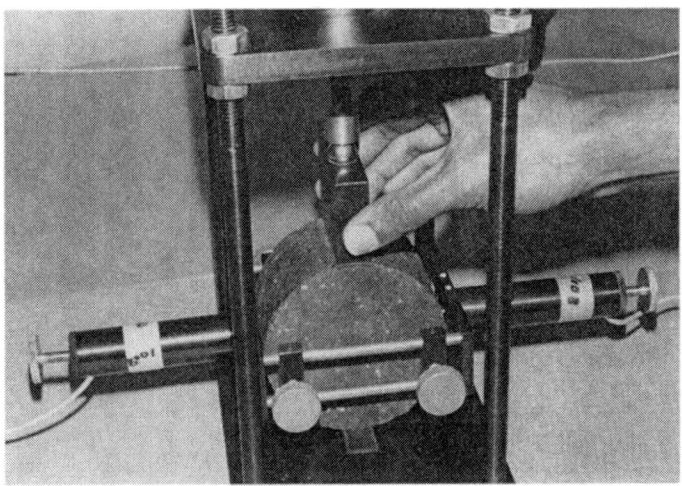

Figure E-14. Seating M_r specimen on loading block.

(13) Calculate the M_r as follows:

S.I. Metric	U.S. Customary

$$M_r = \frac{1000 \, P \, (\nu + 0.2734)}{t\Delta} = 623.4 \, \frac{P}{t\Delta} \qquad M_r = \frac{P \, (\nu + 0.2734)}{t\Delta} = 0.6234 \, \frac{P}{t\Delta}$$

where

- P = dynamic load in kN (lb)
- ν = 0.35 (assumed for Poisson's ratio)
- t = thickness of specimen, mm (in.)
- Δ = deflection in μm (microinches) obtained by multiplying the meter reading by the multiplier.

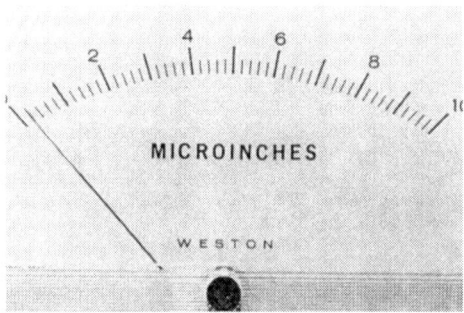

Figure E-15. Adjusting M_r recording meter.

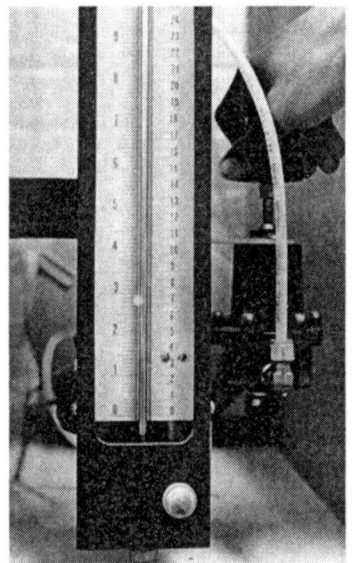

Figure E-16. Adjusting M_r pressure regulator.

Sample Calculations:

<table>
<tr><td align="center">S.I. Metric</td><td align="center">U.S. Customary</td></tr>
<tr><td>

P = 0.333 616 6 kN
t = 63.5 mm
Δ = 3.81 × 10⁻⁷ μm

$$M_r = \frac{623.4 \times 0.3\,336\,166}{63.5 \times 3.81 \times 10^{-7}}$$

M_r = 8 596 383 kPa

</td><td>

P = 75 lb
t = 2.5 in.
Δ = 15 microinches

$$M_r = 0.6234 \times \frac{75}{2.5 \times 15 \times 10^{-6}}$$

M_r = 1,246,800 psi

</td></tr>
</table>

(4) Measure the bulk specific gravity of the specimen according to ASTM D 1188 or D 2726.
(5) Calculate the volume of air and asphalt as follows:

$$V_a = 100 - (V_b + V_{sa})$$

$$V_b = \frac{W_b}{G_b} \times \frac{100}{V_{mb}}$$

$$V_{sa} = \frac{W_s}{G_{sa}} \times \frac{100}{V_{mb}}$$

$$V_{mb} = \frac{W_m}{G_{mb}}$$

$$W_b = \frac{W_e R_e}{100}$$

where

V_a = Volume of air (percent of total mix)

V_b = Volume of asphalt in mix (percent of total mix)

V_{sa} = Volume of aggregate (by apparent specific gravity percent of total mix)

V_{mb} = Bulk volume of compacted mix

W_b = Weight of asphalt

G_b = Specific gravity of asphalt

W_s = Weight of dry aggregate

W_m = Weight of dry compacted mix

G_{mb} = Bulk specific gravity of dry compacted mix

G_{sa} = Aggregate apparent specific gravity

W_e = Weight of emulsified asphalt

R_e = Percent residue of emulsified asphalt, expressed as a whole number.

E.08 MOISTURE EXPOSURE AND STABILITY AND COHESION TESTING.—

a. *General*
Dense-graded mixtures used in the base course or as a temporary wearing surface are evaluated for early strength and fully-cured strength after vacuum saturation.

b. *Vacuum Saturation*
(1) General
This test simulates the effect of prolonged exposure to subsurface water on dense-graded base mixtures.
(2) Equipment
(a) *Vacuum apparatus* shown in Figure E-17.
(b) *Vacuum pump* capable of pulling 100 mm of Hg.

Figure E-17. Vacuum manometer and desiccator.

(3) Procedure
   (a) Record the weight, A_B, of the specimen from the resilient modulus testing.
   (b) Place the specimen into the vacuum apparatus and cover with water (desiccant should be removed from the vacuum apparatus before filling with water).
   (c) Evacuate the desiccator to 100 mm of Hg and hold for one hour.
   (d) Slowly release the vacuum and allow the specimen to soak in water for one hour.
   (e) Remove the specimen, surface dry, reweigh and record this weight as A_W.
   (f) Determine the Stabilometer (R-value) and Cohesiometer (C-value) as subsequently described.
   (g) Dry the entire specimen to constant weight at $110 \pm 5°$ C ($230 \pm 9°$ F), cool to room temperature and weigh. Record this weight as A_D.
   (h) Calculate percent Moisture Pick-up, P, by the specimen during vacuum saturation:

$$P = \frac{A_W - A_B}{A_D} \times 100$$

c. *Resistance R-Value Test*
(1) General
   This test is used to measure the stability or bearing capacity of dense-graded base mixtures at a test temperature of $23 \pm 2.8°$ C ($73 \pm 5°$ F). An early Resistance R-value is determined on "early cure" specimens fabricated by the procedures of Art. E.06 after curing in the mold in a horizontal position for a total of 24 hours at a temperature of $23 \pm 2.8°$ C ($73 \pm 5°$ F) (no moisture exposure). Also, this test is performed on the "fully cured" specimens that are vacuum saturated as described in Art. E.08 (b).

(2) Equipment
   (a) *Hveem stabilometer* and accessories (see Figure E-18).
   (b) *Testing machine*, 222 kN (50,000 lb.) capacity.

(3) Procedure
   (a) Calibrate the displacement of the stabilometer:
      (1) Adjust bronze nut on stabilometer stage base so that the top of the stage is 89 mm (3 1/2 in.) below the bottom of the upper tapered ring. Perform all tests at this stage setting.
      (2) Put a metal dummy specimen in place in the stabilometer. Apply a load of from 0.445 to 0.89 kN (100 to 200 lb.) on the testing machine dial to the dummy specimen to make certain the dummy is held firmly in place. Crank the pump to a pressure of exactly

34.5 kPa (5 psi). Tap the stabilometer dial lightly with fingers in order to be sure the needle is resting on 34.5 kPa (5 psi) pressure. Adjust the turns indicator dial to zero.

Turn pump handle at approximately two turns per second until the stabilometer reads 689 kPa (100 psi). The turns indicator dial should then read 2.00 ± 0.05 turns. If it does not, the air in the cell must be adjusted by means of the rubber bulb, and the displacement measurement must be repeated after each air change until the proper number of turns is obtained. Release horizontal and vertical pressures and remove dummy specimen. The stabilometer is now ready for testing specimens.

(3) Adjust testing machine to give a constant movement of 1.3 mm (0.05 in.) per minute with no load applied. The hydraulic machines must be run several minutes before oil warms sufficiently to maintain a constant speed.

(b) Transfer the specimens to the stabilometer.

(c) Place follower on top of specimen and crank pump to give a horizontal pressure of 34.5 kPa (5 psi). [The 34.5 kPa (5 psi) pressure should be exact; a deviation of only 7 kPa (1 psi) has considerable effect on the final value.]

(d) Start vertical movement of testing machine platen at speed of 1.3 mm (0.05 in.) per minute, and record the stabilometer gauge readings when the vertical forces are 2.224, 4.448, and 8.896 kN (500, 1,000 and 2,000 lb.) total load. [Vertical pressures are 276, 552, and 1,103 kPa (40, 80 and 160 psi).]

Figure E-18. Hveem stabilometer.

(e) Stop vertical loading exactly at 8.896 kN (2,000 lb.) and immediately reduce to a load of 4.448 kN (1,000 lb.)
(f) Turn the displacement pump so that the horizontal pressure is reduced to exactly 34.5 kPa (5 psi). This will result in a further reduction in the vertical loading reading which is normal and for which no compensation is made. Set the turns displacement indicator dial to zero. Turn pump handle at approximately two turns per second until the stabilometer gauge reads 689 kPa (100 psi).

During this operation, the vertical load registered on the testing machine will increase and in some cases exceed the initial 4.448 kN (1,000 lb.) load. As before, these changes in testing machine loading are characteristic and no adjustment or compensation is required.

(g) Record the number of turns indicated on the dial as the displacement of the specimen. The turns indicator dial reads in 0.0254 mm (0.001 in.) and each 2.54 mm (0.1 in.) is equal to one turn. Thus, a reading of 6.35 mm (0.250 in.) indicates that 2.50 turns were made with the displacement pump.

(h) Calculate the Resistance R-value from this formula:

$$R = 100 - \frac{100}{\frac{2.5}{D}\left(\frac{P_v}{P_h} - 1\right) + 1}$$

where P_v = 1 103 kPa (160 psi) vertical pressure
D = turns displacement reading
P_h = horizontal pressure [stabilometer gauge reading for 1 103 kPa (160 psi) vertical pressure]

The chart for determining R-value from stabilometer data (Figure E-19) is normally used to solve the above formula.

(i) Every attempt should have been made to fabricate test specimens having an overall height between 62 mm (2.45 in.) and 65 mm (2.55 in.). However, if for some reason this was not possible, the R-value should be corrected as indicated on the chart for correcting R-values to height of 65 mm (2.5 in.) (Figure E-20.)
(j) Upon completion of R-value determination, remove the specimen and immediately test for cohesiometer value (Art. E.08e).

d. *Stabilometer S-Value*
(1) General
This test measures the stability or bearing capacity of compacted fully cured permanent dense-graded surface mixes (no moisture exposure).
(2) Equipment
(a) *Hveem stabilometer* and accessories (see Figure E-18).
(b) *Testing machine,* 222 kN (50,000 lb.) capacity.

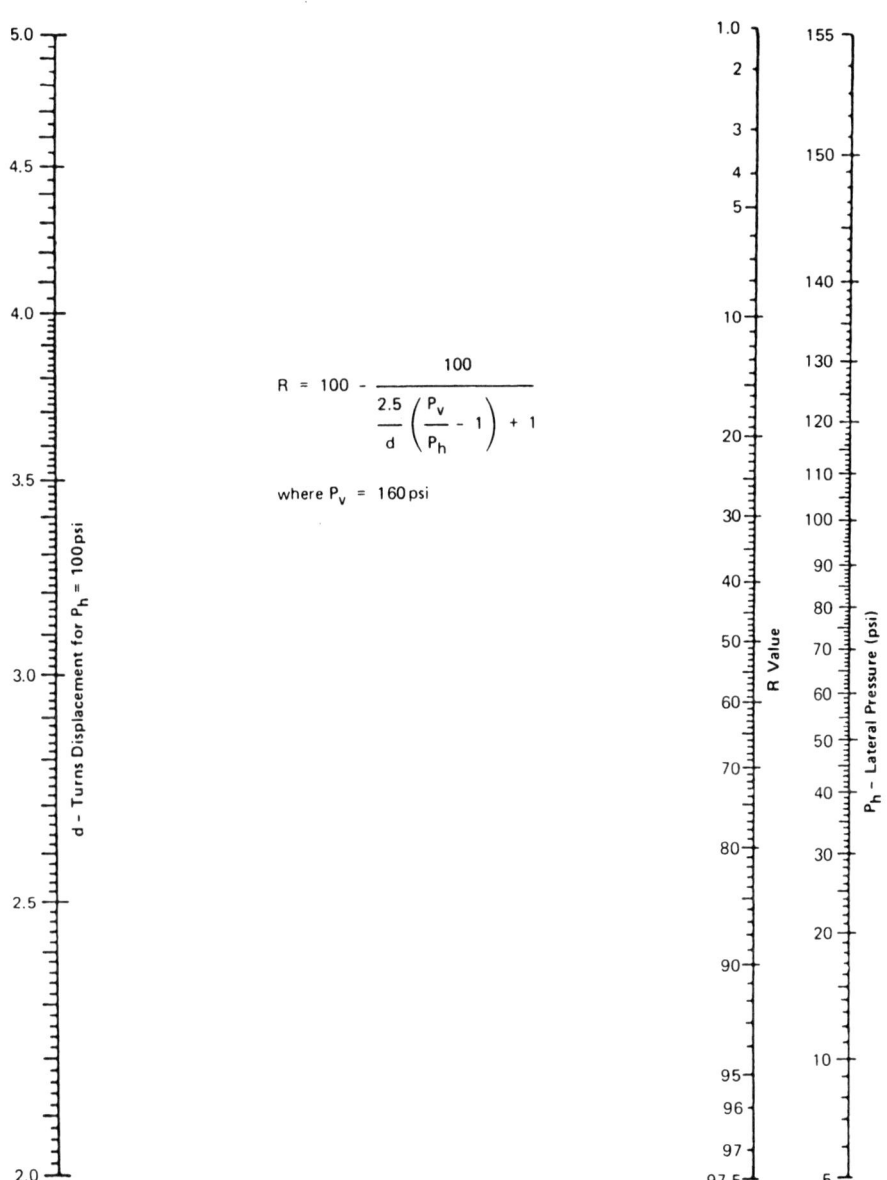

Figure E-19. Chart for determining R-value from stabilometer data; multiply psi by 6.894 757 to obtain kPa.

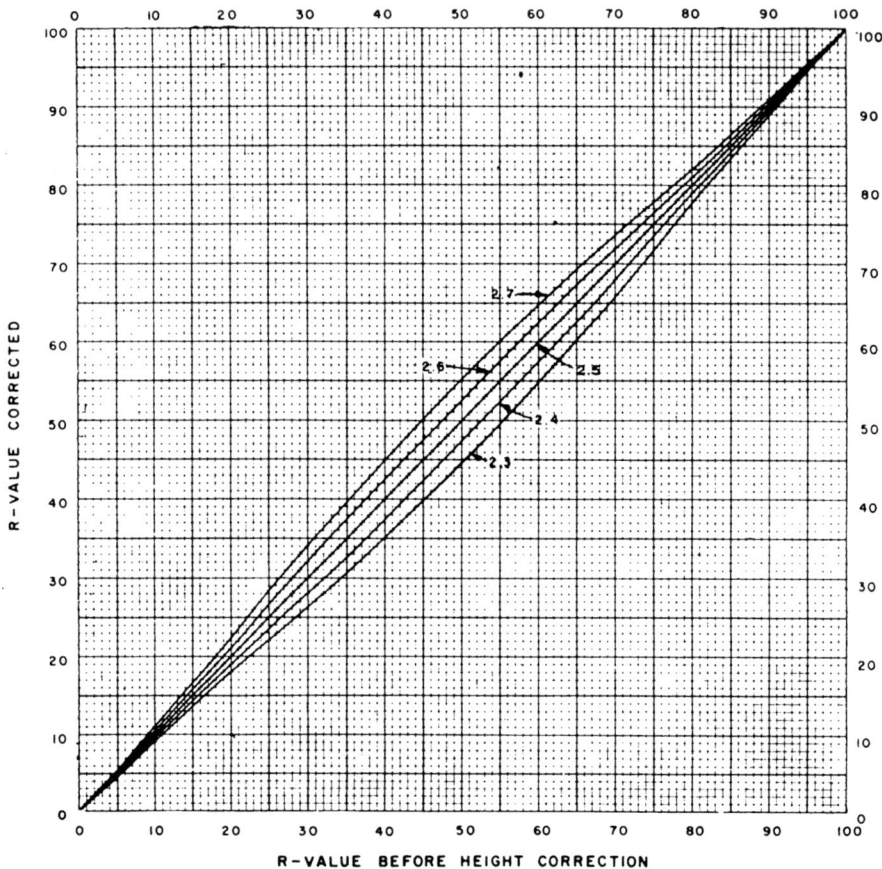

Figure E-20. Chart for correcting R-values to height of 63.5 mm (2.50 in.); multiply in. by 25.4 to obtain mm.

(3) Procedure
  (a) Place the specimens for test in 60 ± 2.8° C (140 ± 5° F) oven for 2 hours before testing.
  (b) Calibrate the displacement of the stabilometer (see Art. E.08d (3)).
  (c) Transfer the specimen to the stabilometer.
  (d) Start vertical movement of testing machine platen at speed of 1.3 mm (0.05 in.) per min., and record the stabilometer gauge readings when the vertical forces are 2.224, 4.448 and each 4.448 kN (500, 1,000 and each 1,000 lb.) thereafter up to 22.24 kN (5,000 lb.).
  (e) Stop vertical loading exactly at 26.69 kN (6,000 lb.) and immediately reduce the load to 4.448 kN (1,000 lb.).
  (f) Turn the displacement pump so that the horizontal pressure is reduced to exactly 34.5 kPa (5 psi). This will result in a further reduction in the vertical loading reading which is normal and for which no compensation is made. Set the turns displacement indicator dial to zero. Turn

pump handle at approximately two turns per second until the stabilometer gauge reads 689 kPa (100 psi).

During this operation the vertical load registered on the testing machine will increase and in some cases exceed the initial 4.448 kN (1,000 lb.) load. As before, these changes in testing machine loading are characteristic and no adjustment or compensation is required.

(g) Record the number of turns indicated on the dial as the displacement of the specimen. The turns indicator dial reads in 0.0254 mm (0.001 in.) and each 2.54 mm (0.1 in.) is equal to one turn. Thus, a reading of 6.35 mm (0.250 in.) indicates that 2.50 turns were made with the displacement pump. This measurement is known as turns displacement of the specimen.

(h) Calculate the stabilometer S-value from the following formula:

$$S = \frac{22.2}{\frac{P_h D_2}{P_v - P_h} + 0.222}$$

where S = stabilometer value
D_2 = displacement on specimen (turns)
P_v = vertical pressure [typically 2 758 kPa (400 psi) = 22.24 kN (5,000 lb.) total load]
P_h = horizontal pressure = stabilometer pressure gauge reading taken at the instant P_v is 2 758 kPa (400 psi) or 22.24 kN (5,000 lb.) total load

(i) Every attempt should have been made to fabricate test specimens having an overall height between 62 mm (2.45 in.) and 65 mm (2.55 in.). However, if for some reason this was not possible, the S-value should be corrected as indicated on the chart for correcting S-values to a height of 64 mm (2.5 in.) (Figure E-21).

(j) Upon completion of S-value determination, remove specimen and immediately test for cohesiometer value (Art. E.08e).

e. *Cohesiometer Test*

(1) General

This test provides a measure of the cohesive resistance or tensile strength of the compacted mixture.

(2) Equipment

(a) *Cohesiometer* (see Figure E-22).

(3) Procedure

(a) Cohesiometer tests are performed on the same specimens previously tested in the stabilometer.

(b) Calibrate cohesiometer device so that the shot (or liquid) will flow into the receiving bucket at the end of 760 mm (30 in.) lever arm at the rate of 1800 ± 20 g per minute.

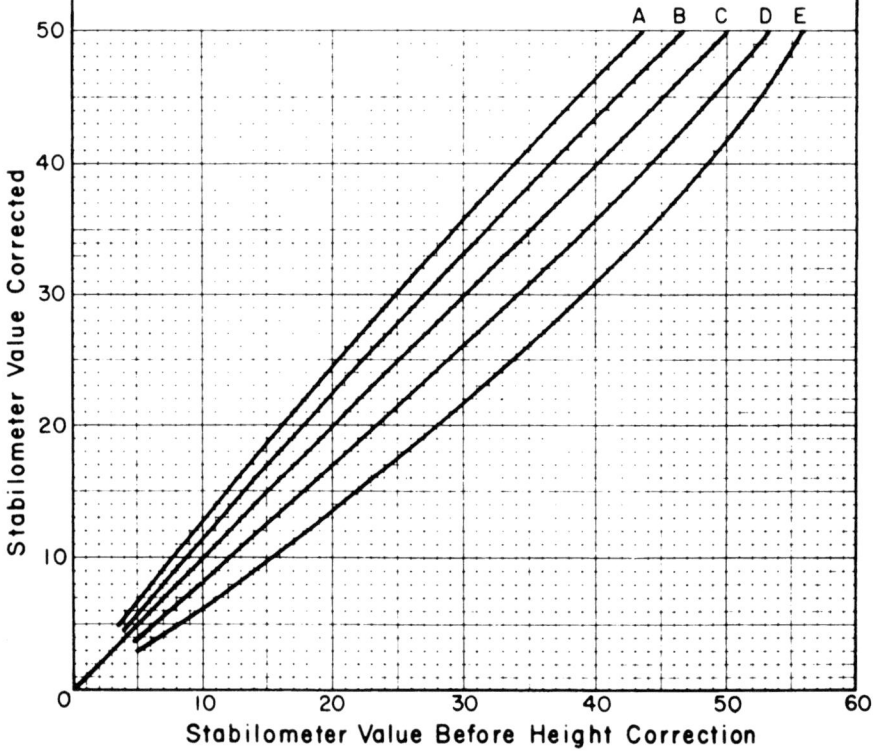

Figure E-21. Chart for correcting stabilometer values to effective specimen height of 64 mm (2.5 in.).

(c) Adjust heating unit in cabinet of cohesiometer device to maintain a temperature of 23 ± 2.8° C (73 ± 5° F) for testing base mixes and 60 ± 2.8° C (140 ± 5° F) for testing surface mixes.
(d) Lock device in position with release pin. Clamp firmly in position, centered, and with top plates parallel with top surface of specimen. Permit temperature in cohesiometer cabinet to equilibrate to the desired temperature before starting the test.
(e) Pull release pin and allow shot flow to continue until the specimen breaks, indicated by a sudden drop of beam.
(f) In the event that the specimen is flexible or ductile rather than brittle, the flow of shot is stopped when the end of the 760 mm (30 in.) beam has lowered 13 mm (1/2 in.) from horizontal.
(g) Weigh shot caught in receiving bucket to the nearest gram and record as *shot weight*.
(h) Calculate cohesiometer value, C, as:

$$C = \frac{L}{W(0.20H + 0.044H^2)}$$

where C = cohesiometer value (g per in. width corrected to a 3 in. height)
L = weight of shot in grams
W = diameter or width of specimen in inches
H = height of specimen in inches

Note: This equation may only be used with U.S. Customary units. Metric units must be converted to obtain the correct cohesiometer value.

Cohesiometer values may also be obtained by multiplying the weight of shot necessary to break the specimen by factors established for various heights of 100 mm (4 in.) diameter (or width) specimens. The factors used are shown in Table E-4.

Table E-4. Multiplying Factors for Cohesiometer Values

| Height | | Factor | Height | | Factor |
mm	(in.)		mm	(in.)	
55.9	(2.20)	.383	63.5	(2.50)	.323
57.2	(2.25)	.372	64.8	(2.55)	.314
58.4	(2.30)	.361	66.0	(2.60)	.306
59.7	(2.35)	.351	67.3	(2.65)	.298
61.0	(2.40)	.341	68.6	(2.70)	.290
62.2	(2.45)	.332	69.8	(2.75)	.283

Example: Assume that it takes 600 g of shot to break a certain specimen which has a 100 mm (4 in.) diameter and 64 mm (2.50 in.) height. Cohesiometer value = 600 x 0.323 = 194.

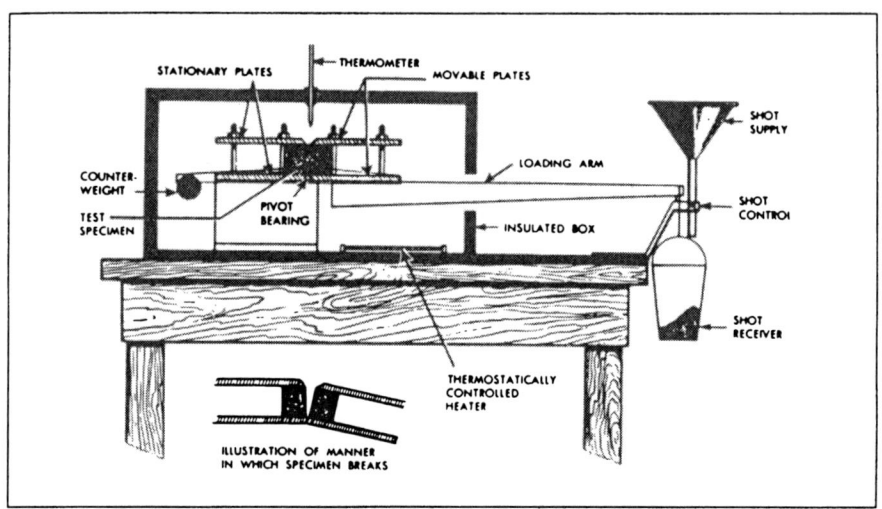

Figure E-22. Diagram showing principal features of the Hveem cohesiometer.

g. *Resistance R_t Value*

Calculate the Resistance R_t Value for base and temporary wearing surfaces according to this formula:

$$R_t = R + 0.05\,C$$

where R = Resistance R-Value [23 ± 2.8° C (73 ± 5° F)]
 C = Cohesiometer C-Value [23 ± 2.8° C (73 ± 5° F)]

E.09 DETERMINATION OF OPTIMUM EMULSIFIED ASPHALT CONTENT.—

An optimum emulsified asphalt content is established on the basis of the best combination of stability, density and maximum resistance to water in addition to meeting the minimum requirements of Table E-5.

Table E-5. Design Criteria for Emulsified Asphalt-Aggregate Mixes

Test Property		Base Mixtures	Surface Mixtures
RESISTANCE R_t-VALUE at $(23 \pm 2.8°C)$ $(73 \pm 5°F)$	Early Cure[a]	70 min.	N.A.
	Fully cured and water immersed[b]	78 min.	N.A.
STABILOMETER S-VALUE at $(60 \pm 2.8°C)$ $(140 \pm 5°F)$		N.A.	30 min.
COHESIOMETER C-VALUE at $(23 \pm 2.8°C)$ $(73 \pm 5°F)$	Early Cure[a]	50 min.	N.A.
	Fully cured and water immersed[b]	100 min.	N.A.
COHEISIOMETER C-VALUE at $(60 \pm 2.8°C)$ $(140 \pm 5°F)$		N.A.	100 min.
AGGREGATE COATING (%)		50 min.	75 min.

[a] Cured in mold for total of 24 hours at temperature of $23 \pm 2.8°C$ $(73 \pm 5°F)$.

[b] Cured in mold for total of 72 hours at temperature of $23 \pm 2.8°C$ $(73 \pm 5°F)$ vacuum desiccated for 4 days followed by water immersion for one hour under vacuum and one hour without vacuum.

N.A. Not Applicable

NOTE: Besides meeting the above requirements, the mix must be reasonably workable (i.e., not too stiff or sloppy).

Appendix F. Marshall Method for Emulsified Asphalt-Aggregate Cold Mixture Design

F.01 SCOPE.—This design for cold-mix emulsified asphalt-aggregate paving mixtures is based on research conducted at the University of Illinois using a modified Marshall method of mix design and a moisture durability test. The method and recommended test criteria are applicable to base course mixtures for low traffic volume pavements containing emulsified asphalt and dense-graded mineral aggregates with maximum sizes of 25 mm (1 in.) or less. This design is recommended for road mixes or plant mixes prepared at ambient temperatures.

F.02 OUTLINE OF METHOD.—
The design procedure involves these major parts:
(1) Aggregate quality tests.
(2) Emulsified asphalt quality tests.
(3) Type and approximate amount of emulsified asphalt.
(4) Water content at mixing and at compaction.
(5) Variation of residual asphalt content.
(6) Selection of optimum asphalt content. The optimum asphalt content is chosen as the percentage of emulsified asphalt at which the paving mixture best satisfies all of the design criteria.

F.03 OBJECTIVE.—
Provide an adequate amount of residual asphalt to economically stabilize granular materials to:
(1) Give required strength or stability to withstand repeated load applications (compressive and flexural) without excessive permanent deformation or fatigue cracking.
(2) Render the mixture sufficiently insensitive to moisture effects.

F.04 AGGREGATES FOR EMULSIFIED ASPHALT MIXES.—
Aggregate properties are the determining factor in many of the choices made concerning the optimum mixture. Thorough testing of the aggregate therefore is necessary. A wide range of materials are suitable for use with emulsified asphalt including crushed stone, rock, gravel, sand, silty sand, sandy gravel, slag, reclaimed aggregate, ore tailings, or other inert materials.

F.05 EMULSIFIED ASPHALTS.—
Selection of emulsified asphalt type and grade for use on a particular project is based in part on the ability of the emulsion to adequately coat the job aggregate. Some factors which affect this selection are:
(1) Aggregate type.
(2) Aggregate gradation and characteristics of the fines.
(3) Anticipated water content of the aggregate.
(4) Availability of water at the construction site.

More than one emulsified asphalt type is often acceptable for a given aggregate, and the selection should be based on mixture properties determined by comparative mixture designs. Additional factors that cannot be evaluated at the time of design of the mixture, but which should be accounted for at the time of construction are:
(1) Anticipated weather at the time of construction.
(2) Type of mixing process.
(3) Construction equipment selected and field procedures used.

F.06 APPROXIMATE AMOUNT OF EMULSIFIED ASPHALT.—

The amount of emulsified asphalt is estimated for trial mixes of dense graded aggregates using the Centrifuge Kerosene Equivalent test (C.K.E.).

The equipment and procedures for running the C.K.E. test are contained in Appendix E, Art. E.05 b and c.

If C.K.E. equipment is not available, an approximation of the emulsified asphalt content for trial mixes can be made:

$$P = (0.05A + 0.1B + 0.5C) \times (0.7)$$

where P = Percent* by weight of emulsified asphalt, based on weight of dry aggregate
A = Percent* of aggregate retained on 2.36 mm (No. 8) sieve
B = Percent* of aggregate passing 2.36 mm (No. 8) sieve and retained on 75 µm (No. 200) sieve
C = Percent* of aggregate passing 75 µm (No. 200) sieve.

*Expressed as a whole number.

F.07 COATING TEST.—

Preliminary evaluation of each emulsified asphalt selected for mixture design is accomplished through a coating test. The trial residual asphalt content as determined in Art. E.06 is combined with the job aggregate, and coating is visually estimated as a percentage of the total area. An emulsified asphalt's ability to coat an aggregate is usually sensitive to the pre-mix water content of the aggregate. This is especially true for aggregates containing a high percentage of material passing a 75 µm (No. 200) sieve, where insufficient pre-mixing water results in balling of the asphalt with the fines and insufficient coating. For this reason, the coating test is performed at varying aggregate water contents. Emulsified asphalts which do not pass the coating test are not considered further. Detailed procedures for the coating test are listed below.

a. *Equipment*
   (1) *Balance,* 5,000 g minimum capacity and accurate to within ± 0.5 g.
   (2) *Laboratory mixing equipment,* preferably mechanized and capable of producing intimate mixtures of the job aggregate, water and emulsified asphalt. Hand mixing, if used, must be sufficiently thorough to uniformly disperse the water and emulsion throughout the aggregate.

(3) *Hot plate*, or 100° ± 5° C (230°± 9° F) oven.
(4) *Pans, metal*, approximately 200 x 355 × 50 mm (8 × 14 × 2 in.).
(5) Supply of *metal kitchen mixing spoons* (approximately 25 mm (10 in.)).
(6) *A one-hundred millilitre glass graduate*.

b. *Procedure*
(1) Obtain representative samples of each emulsified asphalt considered for the project.
(2) Obtain representative samples of the job aggregate or aggregate blend.
(3) Prepare the aggregate by air drying until it is easily separated into sizes using these sieves: 25.0 mm (1 in.), 19.0 mm (3/4 in.), 12.5 mm (1/2 in.), 9.5 mm (3/8 in.), and 4.75 mm (No. 4). Dry until the portion passing the 4.75 mm (No. 4) sieve has a free-flowing consistency. Any suitable means of drying which does not heat the aggregate in excess of 60° C (140° F) or cause degradation of the particles may be used. The aggregate should be stirred frequently to prevent crusting or formation of hard lumps.
(4) Determine the moisture content on a combined sample of the air-dried aggregate according to ASTM Test Method D 2216 and record.
(5) Prepare a sufficient number of batches of the air-dried job aggregate for trial mixes. The batch mass should be approximately 1200 g (oven dry basis). These batches should be prepared by reblending exact fractions of material retained on 25.0 mm (1 in.), 19.0 mm (3/4 in.), 12.5 mm (1/2 in.), 9.5 mm (3/8 in.) and 4.75 mm (No. 4) sieves with material passing 4.75 mm (No. 4) sieve to match the gradation analysis of the whole sample.
(6) Place one batch of aggregate in the mixing bowl of the mechanical mixer. Incorporate a predetermined percentage of water by dry weight of aggregate in excess of the air-dried water content. Water should be added in a thin stream and the aggregate mixed until the water is thoroughly dispersed. (Sixty seconds of mixing time is sufficient.) Select the initial percentage of water by these criteria:
  (a) Medium setting (HFMS, CMS and other solvent containing) asphalt emulsions. Initial trial may be mixed without the addition of any water (i.e., air dry conditions).
  (b) Slow setting (SS and CSS) asphalt emulsions. Often require a higher water content to produce satisfactory mixes; start the coating test at about 3 percent added water.
With aggregates containing clay, the aggregate should be placed in a sealed container for a minimum of 15 hours prior to the addition of emulsified asphalt.
(7) Add the amount of emulsified asphalt (percent by weight of dry aggregate) as determined in Art. F.06. The emulsion should be added in a thin stream to minimize the tendency of the asphalt to ball up with the

fine aggregate. A one-minute mixing process* is usually satisfactory. If hand mixing is used, it should be sufficiently thorough to disperse the asphalt throughout the mixture.

(8) Calculate the free water content of the aggregate at mixing by combining the moisture content of the aggregate as determined in Step (4) with the percentage of water added in Step (6).

Example:
    Water content of [air dried] aggregate = 0.5 percent
    Percentage of water added prior to addition of emulsified asphalt
    = 3.0 percent
    Total pre-mix water before mixing with emulsified asphalt
    = 3.5 percent

(9) Allow the mixtures to air dry with the aid of an electric fan. Prepare a new batch by repeating Steps (6), (7) and (8) with an additional increment of 1 percent water by weight of dry aggregate. Mixes which become soupy or segregate on standing are considered unacceptable. When this occurs proceed to Step (10).

(10) Rate the appearance of the surface dry mixtures by visually estimating the total aggregate surface area that is coated with asphalt. For each pre-mix water content at mixing, record the estimate of the coating as a percentage of the total area. Aggregate coating in excess of 50 percent shall be considered acceptable (see Note 1). If the mixture does not attain 50 percent coating at any water content, the emulsion shall be rejected from further consideration. If the coating appears borderline, the mixture may be evaluated by the full mixture design procedure.

(11) For medium setting (HFMS, CMS and other solvent-containing) asphalt emulsions, use sufficient pre-mix water to give optimum dispersion of the emulsified asphalt. In some cases, excessive pre-mix water may cause stripping of the asphalt from the aggregate. Where this occurs, use only as much water as needed to give at least 50 percent coating. All subsequent mixing shall be done at the water content that produces maximum coating without stripping (see Note 2).

(12) Slow setting (SS and CSS) asphalt emulsion mixtures generally exhibit increased coating as the pre-mix water content is incrementally increased. At some point, sufficient water is available for optimum dispersion of the asphalt and additional increments of water do not improve coating. This result shall be the minimum pre-mix water content required for mixing. All subsequent mixing in the design process shall be done at the minimum pre-mix water content.

---

*Mixing time should be shortened to 30 seconds if segregation of asphalt-fines mixture from coarse aggregate is noticed.

> NOTE 1: It is important to recognize that 100 percent coating common to hot-mixed materials is desirable but not required. Sufficient asphalt to produce 100 percent coating may result in an excessively high asphalt content.
>
> NOTE 2: Some combinations of aggregate and emulsified asphalt are not significantly affected by a variation of water content at mixing. In these cases, mixing may be allowed at or above the optimum water content as determined for compaction.

F.08 OPTIMUM WATER CONTENT AT COMPACTION.—

Mixture properties are closely related to the density of the compacted specimens. Thus, it is necessary to optimize the water content at compaction to maximize the desired mixture properties. This must be done for each combination of emulsified asphalt, type and grade, and aggregate type considered for each project.

The mixture design procedure utilizes standard Marshall specimens in the evaluation of compacted mixture properties. To obtain reliable results, triplicate specimens are prepared for each water content at compaction.

a. *Equipment*

The equipment required for the preparation of test specimens is:
(1) *Scoop*, for batching aggregate.
(2) *Thermometer*, armored, glass or dial type with metal stem, 10° C (50° F) to 65.5° C (150° F).
(3) *Balance, 10 kg capacity*, sensitive to ± 1 g for aggregate and aerating mixtures.
(4) *Balance, 2 kg capacity*, sensitive to ± 0.1 g for compacted specimens and bulk density determination.
(5) *Mixing spoon*, large.
(6) *Spatulas*, small and large.
(7) *Mechanical mixer*, capacity to handle 2500 g.
*(8) *Compaction pedestal* consisting of an 200 × 200 × 460 mm (8 × 8 × 18 in.) wooden post capped with a 305 × 305 × 25 mm (12 × 12 × 1 in.) steel plate. The wooden post should be oak, yellow pine or other wood having a density of 673 to 769 kg/m^3 (42 to 48 lb/ft^3). The wooden post should be secured by four angle brackets to a solid concrete slab. The steel cap should be firmly fastened to the post. The pedestal should be installed so that the post is plumb, the cap level, and the entire assembly is free from movement during compaction.
*(9) *Compaction mold* consisting of base plate, forming mold, and collar extension. The forming mold has an inside diameter of 101.6 mm (4 in.) and height of approximately 76 mm (3 in.); the base plate and collar extension are designed to the interchangeable with either end of the forming mold.

---

*Marshall test apparatus shall conform to requirements of ASTM Test Method D 1559.

*(10) *Compaction hammer* consisting of a flat circular tamping face 98.4 mm (3 7/8 in.) in diameter and equipped with a 4.5 kg (10 lb) mass constructed to obtain a specified 457 mm (18 in.) height of drop.
*(11) *Mold holder*, consisting of spring tension device designed to hold compaction mold in place on compaction pedestal.
(12) *Extrusion jack* or *Arbor press* for extruding compacted specimens from mold.
(13) *Gloves*, welders, for handling hot equipment; gloves, rubber, for removing specimens from oven.
(14) *Marking crayons* for identifying test specimens.
(15) *Pans*, metal, approximately 200 × 355 × 50 mm (8 × 14 × 2 in.) for batching aggregates.
(16) *Oven*, forced draft, capable of maintaining a temperature of 110 ± 2.8° C (230 ± 5° F) for determining moisture contents.

b. *Preparation of Test Specimens*
(1) Number of specimens. Prepare three specimens for each water content at compaction to be evaluated. Generally, three increments of water content one percent apart are sufficient to define the stability (density)/water content at compaction curve.
(2) Preparation of molds and hammer. Thoroughly clean the specimen mold assemblies and the face of the compaction hammer. Place a piece of waxed paper cut to size in the bottom of the mold before placing mixture in the mold.
(3) Preparation of aggregate. Recombine each size fraction of the aggregate to produce a total aggregate mass of 1.2 kg for each batch. Place the pans in a well ventilated area and determine the temperature of the aggregate. The temperature should be adjusted to 22.2 ± 1.7°C (72 ± 3° F) prior to mixing.
(4) Calculations. Four calculations are required for each combination of aggregate and asphalt. They are mass of: aggregate, emulsified asphalt, added pre-mixing water and water loss for compaction. These formulas are used for the calculations:

(a) Mass of air dried aggregate added = $\dfrac{a}{100 - b} \times 100$

(b) Mass of emulsified asphalt = $\dfrac{a \times c}{d}$

(c) Mass of pre-mixing water added = $a \left(f - b - \dfrac{e \times c}{d} \right) / 100$

(d) Mass of water loss for compaction = $a \left(\dfrac{f - g}{100} \right)$

*Marshall test apparatus shall conform to requirements of ASTM Test Method D 1559.

where a = mass of dry aggregate
b = percent water content of air-dried aggregate
c = desired residual asphalt content, percent weight dry aggregate
d = percent residual asphalt in the emulsion
e = percent water in the emulsion = 100 - d
f = percent pre-mix water content at mixing (mass dry aggregate)
g = percent water content at compaction (mass dry aggregate)

Example:
mass of dry aggregate = a = 1200 g
percent water content of air-dried aggregate = b = 0.5 percent
desired residual asphalt content = c = 4.0 percent
percent residual asphalt in the emulsion = d = 65 percent
percent water in the emulsion = e = 35 percent
percent pre-mix water content at mixing = f = 5.0 percent
percent water content at compaction = g = 3.5 percent

(a) Mass of air-dried aggregate added = $\dfrac{1200}{100 - 0.5} \times 100 = 1206$ g.

(b) Mass of emulsified asphalt = $\dfrac{1200 \times 4.0}{65} = 74$ g.

(c) Mass of added pre-mixing water = $1200 \left(5.0 - 0.5 - \dfrac{35 \times 4.0}{65}\right)/100 = 28$ g.

(d) Mass of water loss for compaction = $1200 \left(\dfrac{5.0 - 3.5}{100}\right) = 18$ g.

Appropriate input values for the previous formulas are discussed in subsequent sections.

(5) Addition of pre-mixing water. Place the air dried aggregate in the mechanical mixer. Calculate the total amount of free water that needs to be added to achieve the optimum pre-mixing water as determined in the coating test. (Art. F.07).

Measure the volume of added water in a graduated cylinder. The temperature of the water shall be 22.2 ± 1.7° C (72 ± 3° F). Add the water in a slow stream and mix for 1 ± 0.5 minutes or until the water is thoroughly dispersed throughout the aggregate. For aggregates containing clay the material shall be placed in a sealed container for a minimum of 15 hours (see Note below). Determine the mass of emulsified asphalt container and record. Add the emulsified asphalt to the moistened aggregate in a thin stream as the material is mixing. Reweigh the emulsified asphalt container periodically to ensure the required mass of emulsified asphalt is not exceeded. A one minute mixing time using a mechanical mixer should be sufficient. Excessive mixing tends to strip the asphalt from the aggregate and should be avoided.

(6) Aeration to reduce the water content of the mixture to get maximum density. If the desired water content at compaction differs from the optimum mixing water content, aeration is required. Distribute the mixture in the pan such that the depth does not exceed 25 millimetres (1 in.). Record the mass of the mixture and pan. The required loss to reach the desired compaction water content is calculated by the equation in F.08b (4)(d). The required loss is subtracted from the recorded mass of mixture and pan and that mass recorded. A fan may be used to aerate the mixture. Stir and weigh the mixture every 10 ± 0.5 minutes until the calculated required water loss is complete. The mixture is now ready for compaction.

> **NOTE:** If coating of the aggregate is not sensitive to the water content at mixing as determined in the coating test (Art. F.07), the aggregate may be mixed at the desired water content at compaction, emulsion added, and the mixture compacted immediately.

(7) Compaction of specimens. For specimens to be tested in the modified Marshall stability test use standard Marshall forming molds. Assemble the base plate, Marshall forming mold, and collar extension. Cover the base plate with a piece of waxed paper cut to size and place mixture in the mold assembly (see Note below). Spade the mixture with a small spatula 15 times around the perimeter and 10 times over the interior. Place a second piece of waxed paper cut to size over the top of the mixture. Repeat this process for the remaining mold assemblies.

Place the first mold assembly on the compaction pedestal in the mold holder and apply 50 blows with the compaction hammer. Remove the collar and base plate, reverse the mold and reassemble. Apply the same number of compaction blows to the face of the reversed specimen. Repeat the process for the remaining mold assemblies. Remove the collars, base plates, and paper from all specimens. Specimens are now ready for curing.

(8) Curing of specimens. Cure in the mold for 1 day at room temperature, with molds on their edge for equal ventilation on both ends, and then extrude. After extrusion, the bulk specific gravities of the specimens are determined by displacement in water (ASTM D 1188 or D 2726).

(9) A plot is made of dry density versus fluids content at compaction. The fluids content resulting in the highest density is optimum for compaction. (See "Mix Design Calculations for use with Table F-1").

If further definition of results is required, batches with an additional water content at compaction may be prepared. The optimum water content at compaction shall be used on all subsequent compaction regardless of the residual asphalt content.

NOTE: Generally it is desirable to prepare a single trial specimen for each type of aggregate considered for the job prior to compacting the test specimens. Should the height of the extruded trial specimen fall outside the limits of 63.5 ± 6 mm (2.5 ± 0.25 in.), the amount of mixture may be adjusted as follows:

$$\text{Adjusted mass of aggregate per specimen} = \frac{63.5 \text{ (mass of aggregate used)}}{\text{(specimen height (mm) obtained)}}$$

or for U.S. Customary Units:

$$\text{Adjusted weight of aggregate} = \frac{2.5 \text{ (weight of aggregate used)}}{\text{(specimen height (in.) obtained)}}$$

F.09 VARIATION OF RESIDUAL ASPHALT CONTENT.—

In determining the optimum residual asphalt content for a particular aggregate and emulsified asphalt combination, a series of test specimens are prepared over a range of residual asphalt contents, using the previously established optimum water contents for mixing and compaction.

Test mixtures are prepared in one percent increments of residual asphalt content with two increments on either side of the trial asphalt content determined in Art. F.06. If further definition of test results is required, increments farther away from the trial residual asphalt content are prepared.

a. *Equipment*
The equipment required for preparation of specimens is listed under Art. F.08a.

b. *Preparation of Specimens*
Use the Procedure for Preparation of Specimens listed in Art F.08b. Additional instructions and clarifications presented below correspond to the appropriate sections of F.08b.
   (1) Number of specimens. Prepare six specimens for each residual asphalt content.
   (2) Preparation of molds and hammer. No change.
   (3) Preparation of aggregate. Use a total aggregate mass of 1,200 g for each specimen batch.
   (4) Calculations. No change.
   (5) Addition of mixing water. Note that the optimum total compaction water is used for all asphalt contents. As the residual asphalt content increases, the amount of water contributed by the emulsion increases. Thus, the amount of

pre-mix water added will be reduced as the residual asphalt content is increased. Vary the residual asphalt content on successive batches to yield five one-percent increments (the trial residual asphalt content and one and two percent increments both sides of the trial).
(6) Aeration to reduce the water content of the mixtures. No change.
(7) Compaction of specimens. No change.
(8) Curing of specimens. Cure in the mold for 1 day at room temperature, extrude and cure for 1 day out of mold in oven at 38° C (100° F).

F.10 TEST PROCEDURE.—

To complete the mix design, tests and analysis are made from data obtained from the compacted specimens:
(a) Bulk Specific Gravity.
(b) Modified Marshall Stability and Flow of Dry Specimens at 22.2 ± 1.1° C (72 ± 2° F).
(c) Soaked Stability and Flow after vacuum saturation and immersion.
(d) Density and Voids Analysis.
(e) Moisture Absorption.

Table F-1 is a detailed data sheet that can be used to record pertinent data and perform calculations.

a. *Equipment*

The equipment required for the testing of the 102mm (4 in.) diameter x 64 mm (2 1/2 in.) height specimens is:
(1) *Marshall Testing Machine,* a compression testing device, conforming to ASTM Test Method D 1559. It is designed to apply loads of test specimens through semicircular testing heads at a constant rate of strain of 50.8mm (2 in.) per minute. It is equipped with a calibrated proving ring for determining the applied testing load, a Marshall stability testing head for use in testing the specimen and a Marshall flow meter for determining the amount of strain at the maximum load for the test. A universal testing machine equipped with suitable load and deformation indicating devices may be used instead of the Marshall testing frame.
(2) *Water Bath.* At least 610 x 915 x 155mm (24 x 36 x 6 in.) and thermostatically controlled at 22.2 ± 1.1° C (72 ± 2° F).
(3) *Pans,* either 229 x 229 mm (9 x 9 in.) or 254 mm (10 in.) in diameter and 25mm (1 in.) deep capable of containing failed specimens for moisture content determination.
(4) *Balances,* 1500g capacity equipped for bulk density determination.
(5) *Towel,* cloth for drying samples during bulk density determination.
(6) *Vacuum pump,* vacuum desiccator and manometer. (See Appendix E, Art. E.08.)

b. *Bulk Specific Gravity Determination*

The method used for determination is ASTM D 2726, "Bulk Specific Gravity and Density of Compacted Bituminous Mixtures Using Saturated Surface-Dry Specimens" or ASTM D 1188, "Bulk Specific Gravity and Density of Compacted Bituminous Mixtures Using Paraffin-Coated Specimens."

Table F-1. Emulsified Asphalt Mixture Data Sheet (Use for specimens containing a single residual asphalt content)

ASPHALT		AGGREGATE	
Type & Grade		Source Id.	
Asphalt in Emulsion	%	Type	
Asphalt Spec. Gra. (B)		Apparent Spec. Grav. (C)	
Residual Asphalt in Mixture (A)	%		
MIXING AND COMPACTION		TESTING	
Total Mix Water	%	Dry Spec. Test Date	
Added Mix Water	g	Rotate Soak Spec. Date	
Water at Compaction	%	Soak Spec. Test Date	
Compaction Date			

COMPACTED SPECIMEN DATA	Dry			Soaked		
	1	2	3	4	5	6
Bulk Density						
Mass in Air (D)						
Mass in Water (E)						
Mass SSD (F)						
BSG - compacted mix (G)						
Dry BSG - compacted mix						
Thickness						
Stability						
Dial						
Load						
Adjusted Stability (S)						
Flow						
Moisture Content						
Mass of specimen (H)						
Mass of oven-dry specimen (I)						
Tare (J)						
Moisture content (K)						
Moisture absorbed						
Maximum Total Voids - %						

MIX DESIGN CALCULATIONS
FOR USE WITH TABLE F-1

$$G \text{ (bulk specific gravity)} = \frac{D}{F - E}$$

$$G_d \text{ (dry bulk specific gravity)} = G \times \frac{(100 + A)}{(100 + A + K)}$$

Dry density, Kg/m^2 = 1,000 × G_d (lb/ft^3 = G_d × 62.4)

$$K \text{ (water content at testing), \%} = \frac{\text{mass of water, g}}{\text{mass of dry mixture, g}} \times (100 + A)$$

$$\text{VMA, \%} = \left[\left(\frac{100 + A + K}{G} - \frac{100}{C} \right) \div \left(\frac{100 + A + K}{G} \right) \right] \times 100$$

$$V \text{ (total voids), \%} = \left[\left(\frac{100 + A + K}{G} - \frac{100}{c} - \frac{A}{B} \right) \div \left(\frac{100 + A + K}{G} \right) \right] \times 100$$

$$\text{Air Voids, \%} = V - \left[\left(\frac{K \times 100}{L} \right) \div \left(\frac{100 + A + K}{G} \right) \right]$$

$$\text{Percent stability loss} = \frac{\frac{S_1 + S_2 + S_3}{3} - \frac{S_4 + S_5 + S_6}{3}}{\frac{S_1 + S_2 + S_3}{3}} \times 100$$

$$\text{Moisture absorbed} = \frac{K_1 + K_2 + K_3}{3} - \frac{K_4 + K_5 + K_6}{3}$$

where
- D = mass of specimen in air, g;
- E = mass of specimen in water, g;
- F = mass of specimen in saturated surface-dry (SSD) condition, g;
- A = asphalt residue as percent of dry aggregate mass;
- B = specific gravity of asphalt;
- C = apparent specific gravity of aggregate;
- L = specific gravity of water;
- S = adjusted stability

Note: Letters A through S refer to identical letters in parentheses in Table F-1

c. Modified Stability and Flow Tests

After determining the bulk specific gravity on six cured specimens, test three of them for stability and flow: (In the selection of test specimens, the three retained for water conditioning should have the same average density as the three tested dry.)

(1) Determine mass of cured specimens and record in column "Mass of Specimen (H)"

(2) Thoroughly clean the guide rods and inside surfaces of the test heads prior to making the test, and lubricate the guide rods so that the upper test head slides freely over them. The testing head temperature is maintained between 21.1 and 23.3° C (70 and 74° F) using a water bath when required. Check the load measuring device for "zero" adjustment.

(3) Place one of the three specimens on the lower testing head into position and center complete assembly in the loading device. Place the flow meter over marked guide rod.

(4) Apply testing load to specimen at constant rate of deformation of 50.8 mm (2 in.) per minute until failure is obtained. The total number of Newtons (lb) required to produce failure of the specimen at 22.2 ± 1.1°C (72 ± 2° F) shall be recorded as its adjusted Marshall stability values (S).

(5) While the stability test is in progress, hold the flow meter firmly in position over the guide rod and remove it the instant the maximum load starts to decrease. Note and record the indicated flow value in units of 0.25mm (0.01 in.)

(6) Place the failed specimens in tared pans, taking care to make sure all of the specimen is put into the pan. The specimens are broken up and put in an oven at 93 ± 6° C (200 ± 10° F). The specimens are removed after 24 hours, reweighed, and the masses recorded under the heading "Mass of Oven Dried Specimen (I)." The mass of the water is corrected by subtracting the mass of water absorbed during bulk specific gravity determination. The mass of the water absorbed can be determined by subtracting the mass of the dry specimens from the mass of the saturated surface-dry specimen. From the data obtained above, a moisture content at testing is determined.

d. Soaked Stability and Flow Tests

After testing three of the six cured specimens for each residual asphalt content, the remaining three specimens are subjected to vacuum saturation and immersion.

(1) Each specimen is separately placed into the vacuum apparatus [See Appendix E, Art. E.08b (2)], and covered with water (desiccant should be removed from the vacuum apparatus before filling with water).
(2) Evacuate the desiccator to 100mm of Hg and hold for one hour.
(3) Slowly release the vacuum and allow specimen to soak in water for one hour.
(4) The specimens are then tested in modified Marshall stability and moisture content determination as outlined in Art. F.10c.

e. Density and Voids Analysis
A density and voids analysis is conducted:
(1) Determine each unit weight in kg/m^3 by multiplying the bulk specific gravity by 1000 (for unit weight in lb/ft^3 multiply by 62.4).
(2) After determining water content at testing, aggregate apparent specific gravity, asphalt specific gravity, and mix bulk specific gravity, voids are calculated as shown under "Mix Design Calculations for use with Table F-1."

Voids are calculated for each specimen. Any values that are more than 50 percent from the average of the three specimens should not be used.

F.11 INTERPRETATION OF TEST DATA
The stability, flow, voids, bulk density, and moisture content data are prepared:
(1) Measured stability values for specimens that depart from the standard 63.5 mm (2-1/2 in.) thickness shall be converted to an equivalent 63.5mm (2-1/2 in.) value by means of a conversion factor. Applicable correlation ratios to convert the measured stability values are set forth in Table F-2. Note that the conversion may be made on the basis of either measured thickness or measured volume.

Table F-2. Stability Correlation Ratios

Volume of Specimen cm^3	Approximate Thickness of Specimen		Correlation Ratio
	mm	in.	
457 to 470	57.2	2 1/4	1.19
471 to 482	58.7	2 5/16	1.14
483 to 495	60.3	2 3/8	1.09
496 to 508	61.9	2 7/16	1.04
509 to 522	63.5	2 1/2	1.00
523 to 535	64.0	2 9/16	0.96
536 to 546	65.1	2 5/8	0.93
547 to 559	66.7	2 11/16	0.89
560 to 573	68.3	2 3/4	0.86

NOTES:
1. The measured stability of a specimen multiplied by the ratio for the thickness of the specimen equals the corrected stability for a 63.5 mm (2 1/2-in.) specimen.
2. Volume-thickness relationship is based on a specimen diameter of 101.6 mm (4 in.).

(2) Average the flow values and the converted stability values for all specimens of a given asphalt content. Values that are obviously in error shall not be included in the average.
(3) Prepare a separate graphical plot for these factors as illustrated in Figure F-1:
   (a) Dry and soaked stability versus residual asphalt content.
   (b) Percent stability loss [calculated by (dry stability-soaked stability) × 100/dry stability] versus residual asphalt content
   (c) Dry bulk density (corrected for moisture) versus residual asphalt content.
   (d) Percent moisture absorbed versus residual asphalt content.
   (e) Percent total voids (air plus moisture) versus residual asphalt content.

In each graphical plot, connect the data with a smooth curve that provides the best fit for all values.

a. *Trends and Relations of Test Data*

The test property curves as previously plotted have been found to vary considerably between aggregate types and gradations, but typical curves are shown in Figure F-1. General trends are:
(1) Soaked stability will generally show a peak at a particular residual asphalt content while dry stability will generally show a continually decreasing curve with increasing residual asphalt content. Some mixes may show a continual increase in soaked stability over the range of asphalt content evaluated, which indicates the increased beneficial effect of additional asphalt content on soaked stability.
(2) Percent loss of stability generally decreases as residual asphalt content increases.
(3) Dry bulk density usually peaks at a particular residual asphalt content.
(4) Percent moisture absorbed during the soak test decreases with increased residual asphalt content.
(5) Percent total voids (air plus moisture) decreases as residual asphalt content increases.

b. *Determination of Optimum Asphalt Content*
(1) Mixture must provide an adequate stability when tested in a "soaked" condition to provide adequate resistance to traffic load during wet seasons.
(2) The percent loss of stability of the mixture when tested "soaked" as opposed to "dry" should not be excessive. A high loss is indicative of the mixture having high moisture susceptibility and may cause disintegration during wet seasons.
(3) The total voids within the mixture should be within an acceptable range to prevent either excessive permanent deformation and moisture absorption (for too high void content), or bleeding of the residual asphalt from the mixture (for a low void content).

(4) Moisture absorption into the mixture should not be excessive to minimize the potential of stripping or weakening the bond between residual asphalt and aggregate.
(5) Residual asphalt should provide adequate coating of the aggregate and should be resistant to stripping or abrasion.

The optimum residual asphalt content for the paving mixture is determined from the data obtained as presented. The optimum residual asphalt content is chosen that provides maximum soaked stability, but is adjusted either up or down depending on moisture absorption, percent loss of stability, total voids, and coating of aggregates. Design criteria for each of these values is given in Table F-3. If the residual asphalt content at the peak of the soaked stability curve provides for adequate moisture absorption, percent loss of stability, total voids, and aggregate coating, it is selected as the optimum asphalt content. This value must meet minimum stability requirements, however, as given in Table F-3, or the mix is rejected. If one or more criteria cannot be met, the mix should be considered inadequate.

If no peak in residual asphalt content versus soaked stability or other properties is developed, the optimum emulsion content should be established based on the best combinations of such properties as Marshall stability of both cured and immersed specimens, percent stability loss and dry density, with particular attention to the effects of water on specimen properties.

Table F-3. Emulsified Asphalt-Aggregate Mixture Design Criteria

Test Property	Minimum	Maximum
Stability, N (lb) at 22.2° C (72° F) Paving Mixtures	2224 (500)	—
Percent Stability Loss After vacuum saturation and immersion	—	50
Aggregate Coating (percent)	50	—

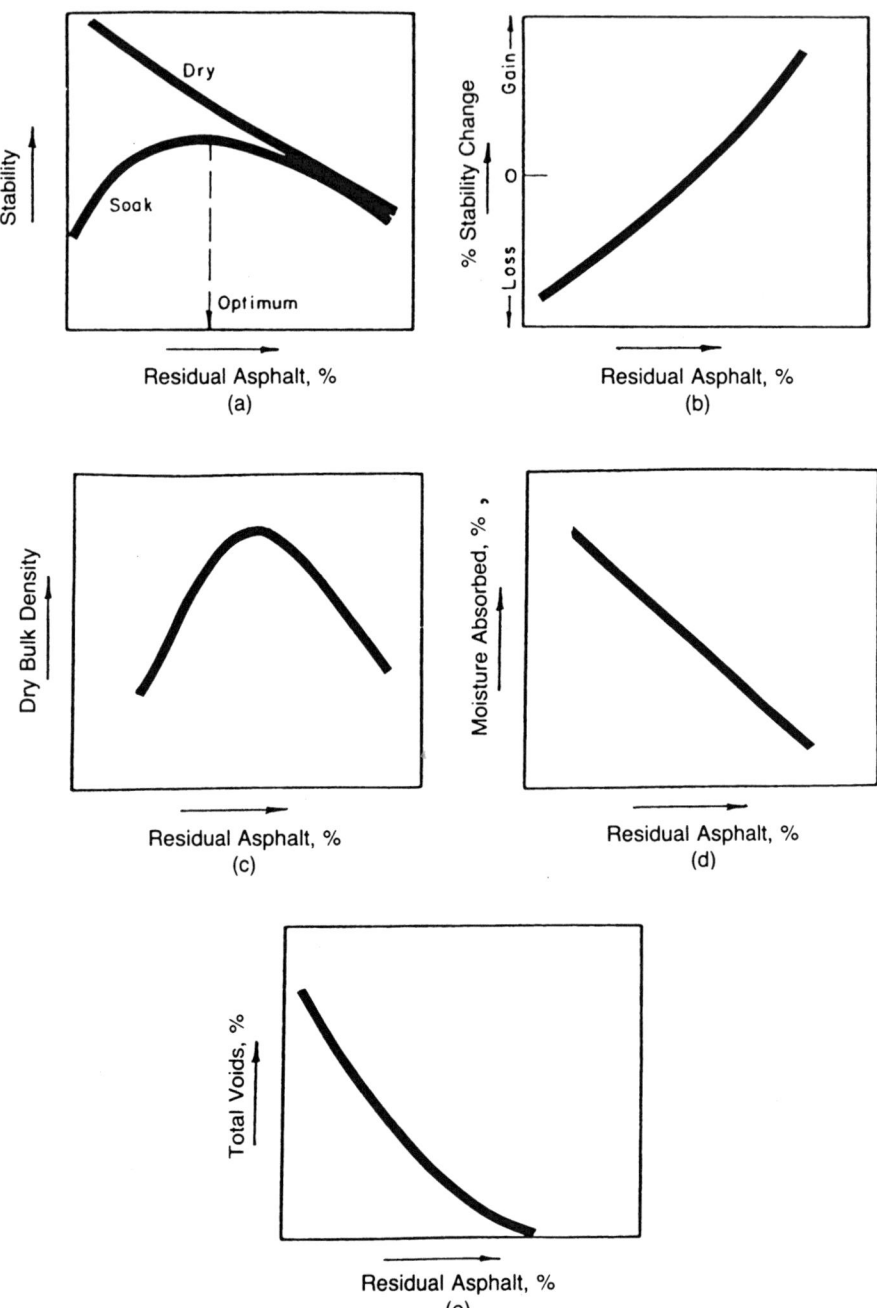

Figure F-1. Typical emulsified asphalt-aggregate mixture design plots.

Appendix G. Modified Hveem Method for Cutback Asphalt-Aggregate Cold Mixture Design

A. General

G.01 APPLICATION.—The mix design method for cutback asphalt mixtures presented here utilizes test procedures from the Hveem Method of Mix Design for hot-mix asphalt paving as described in *Mix Design Methods for Asphalt Concrete and Other Hot-Mix Types,* MS-2, Asphalt Institute, and California Division of Highways test methods and criteria or modifications of these test methods and criteria. This method is applicable to paving mixtures containing cutback asphalt of the medium curing (MC) and slow curing (SC) types and mineral aggregates with maximum sizes not exceeding 25 mm (one inch) in diameter. This includes road mixes and plant mixes. Aggregate gradation requiurements specified for mix designations of 1 in., 3/4 in., 1/2 in., 3/8 in. and No. 4 nominal maximum size, (ASTM D 3515), are suitable for cutback asphalt mixes. Aggregates suitable for cutback asphalt mixtures should have a minimum sand equivalent value (ASTM D 2419) of 35, and K_c and K_f factors obtained from the Centrifuge Kerosene Equivalent test should not exceed 1.8.

G.02 OUTLINE OF METHOD.—The procedure for the Hveem method as presented herein starts with the preparation of test specimens. Preliminary to this operation it is required that:
  (a) the materials proposed for use meet the requirements of the project specifications;
  (b) aggregate blend combinations are determined which meet the gradation requirements of the project specifications; and
  (c) an ample supply of aggregates is dried and sized into fractions.

These requirements are matters of routine testing, specifications, and laboratory technique, which must be considered but are not unique to any particular mix design method. (Refer to MS-2 for additional information on the preparation and analysis of aggregates.) It should be noted, however, that the maximum size aggregates used in the test mixes should not exceed 25 mm (1 in.). In the event the specifications for the paving mix being considered require aggregate sizes greater than 25 mm (1 in.), oversize rock up to 25 percent may be screened out. However, this can have an effect on the magnitude of the stabilometer values depending on the size and amount of the larger aggregate pieces.

The Hveem method uses standard test specimens of 63.5 mm (2 1/2 in.) height by 101.6 mm (4 in.) diameter; these are prepared using a specified procedure for heating, mixing, and compacting the asphalt-aggregate mixtures. The principal features of the Hveem method of mix design are the Centrifuge Kerosene Equivalent (C.K.E.) test on the aggregates to estimate the asphalt requirements of the mix, followed by a stabilometer test, a cohesiometer test, a swell test, and a moisture vapor susceptibility test on specimens of the compacted paving mixtures. The stabilometer test utilizes a special triaxial-type testing cell for measuring the resistance of the compacted mix to lateral displacement under vertical loading. The cohesiometer test measures the cohesive or tensile resistance of the compacted mix. The swell test measures the resistance of the mix to the action of water. The test specimens are maintained at 60° C (140° F) for both the stability and cohesiometer tests, but the swell test is performed at room temperature.

Samples of aggregate and cutback asphalt proposed for the construction and meeting requirements of the project specifications are obtained for the mix design tests. The design procedures include these steps:

(1) Aggregate samples are analyzed and prepared according to MS-2.
(2) Estimated asphalt content for the cutback asphalt paving is determined by the Centrifuge Kerosene Equivalent Method according to Articles G.03 through G.08.
(3) Cutback asphalt paving mixtures are prepared for stabilometer test and swell test according to Articles G.09 through G.13 and moisture vapor susceptibility test according to Article G.13.
(4) Swell, stabilometer, and cohesiometer tests are performed on the specimens according to Articles G.16, G.17, and G.19, respectively. Moisture vapor susceptibility test is performed according to Article G.18.
(5) Suitability of the mix design is determined on the basis of whether stabilometer, swell, and moisture vapor susceptibility test values meet recommended requirements in Article G.21.

B. Estimated Asphalt Content by Centrifuge Kerosene Equivalent Method

G.03 GENERAL.—The first step in the Hveem method of mix design is to determine the "estimated optimum" asphalt content by the Centrifuge Kerosene Equivalent method. With a calculated surface area and the factors obtained by the C.K.E. method for a particular aggregate or blend of aggregates, the estimated optimum asphalt content is determined by using a series of charts. These charts are presented in Appendix E, Article E.05.

G.04 EQUIPMENT.—The equipment and materials required for determining the estimated optimum asphalt content are as contained in Appendix E, Article E.05b(2).

G.05 SURFACE AREA.—The gradation of the aggregate or blend of aggregates employed in the mix is used to calculate the surface area of the aggregates. This calculation consists of multiplying the total percent passing each sieve size by a "surface-area factor" as set forth in Appendix E, Article E.05b(3), and Table E-2. Add the products thus obtained and the total will represent the equivalent surface area of the sample in terms of m²/kg (ft²/lb). It is important to note that all the surface-area factors must be used in the calculation. Also, if a different series of sieves is used, different surface-area factors are necessary.

G.06 C.K.E. PROCEDURE.—(See Appendix E, Article E.05b(4).)

G.07 SURFACE CAPACITY TEST FOR COARSE AGGREGATE.—(See Appendix E, Article E.05b(5).)

G.08 ESTIMATED OPTIMUM ASPHALT CONTENT.—
(a) Using the C.K.E. value obtained and the chart in Appendix E, Figure E-2, determine the value K_f (surface constant for fine material).
(b) Using the percent oil retained and the chart Appendix E, Figure E-3, determine the value K_c (surface constant for coarse material).
(c) Using the values obtained for K_f and K_c, and chart in Appendix E, Figure E-4, determine the value K_m (surface constant for combined aggregate).
(d) The next step is to determine the estimated asphalt content for the mix based on a cutback asphalt with a Saybolt-Furol viscosity range at 60° C (140° F) of 100 to 500 seconds; either of two methods is used, referred to hereinafter as Case I and Case II.
(e) Case I applies to mixes in which the coarse and fine aggregates have similar surface and absorption characteristics. (K_f is approximately equal to K_c). This method, however, is generally used as a field method and is utilized in the central laboratory only when the sample consists mainly (85 percent or more) of material passing the 4.75 mm (No. 4) sieve. The asphalt content based on liquid asphalt is determined by using the percent of aggregate passing the 4.75 mm (No. 4) sieve, the C.K.E. value adjusted for specific gravity (use "apparent" specific gravities) and scales A, B, and C on the alignment chart Appendix E, Figure E-5.
(f) Case II applies to mixes in which the coarse and fine aggregates have dissimilar surface and absorptive characteristics. (K_f and K_c are markedly different.) However, in the central laboratory this method should be used in all cases where the percent passing the 4.75 mm (No. 4) sieve is less than 85 percent. The asphalt content based on a liquid asphalt is determined using the calculated surface area of the sample (see Article G.05), the average "apparent" specific gravity of the aggregates, the surface constant K_m, and scales D to B on the alignment chart in Figure E-5.
(g) Determine the asphalt content (Figure G-1) for mix, (corrected for grade of cutback asphalt to be employed), using the surface area of the sample, the grade of asphalt, and the oil ratio from Figure E-5.

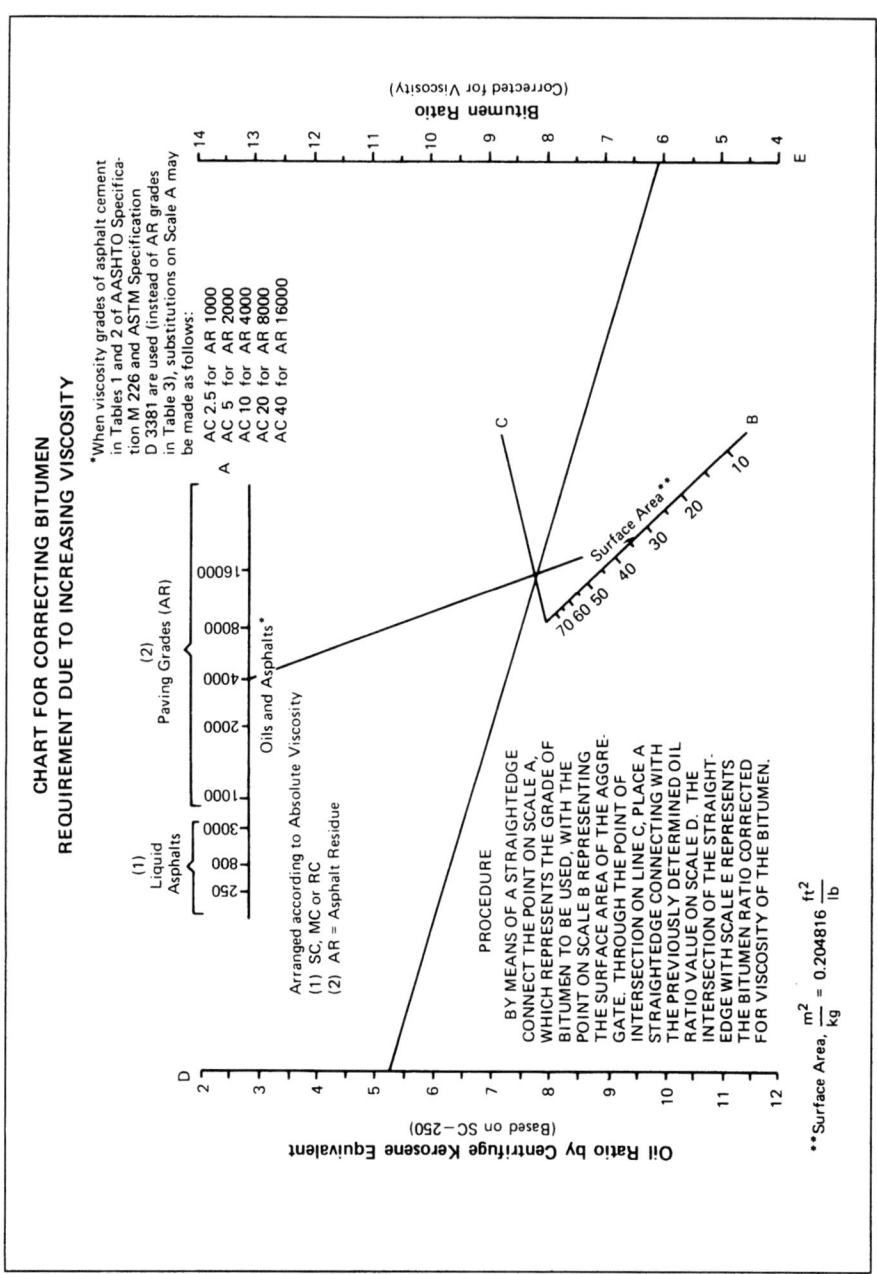

Figure G-1. Chart for correcting asphalt requirement due to increasing viscosity of asphalt, Hveem Method of Design
(Chart courtesy of California Department of Transportation)

EXAMPLE

This demonstrates the use of this procedure. Assume these conditions apply to a paving mix using MC-800 asphalt:

Specific Gravity, coarse = 2.45
Specific Gravity, fine = 2.64
Percent Passing 4.75 mm (No. 4) = 45

$$\text{Avg. Sp. Gr.} = \frac{100}{\frac{55}{2.45} + \frac{45}{2.64}} = 2.53$$

Surface Area of Aggregate Grading = 6.6 m²/kg (32.4 ft²/lb)
C.K.E. = 5.6
Percent Oil Retained, coarse = 1.9
    (corrected for specific gravity, this value is 1.7 percent. See Figure E-3, Appendix E)

From Figure E-2 determine K_f as 1.25.
From Figure E-3 determine K_c as 0.8.
From Figure E-4 determine K_m as 1.10.
From Figure E-5 determine the oil ratio for liquid asphalt as 4.6 percent. (Case II applies here since K_c and K_f are markedly different.)
From Figure G-1 determine optimum asphalt content (oil ratio) for MC-800 asphalt as 4.8 percent by weight of dry aggregate.

C. Preparation of Test Specimens

G.09 GENERAL.—In designing a paving mix by the Hveem method a series of stabilometer test specimens is prepared for a range of asphalt contents, both above and below the "estimated optimum" asphalt content indicated by the C.K.E. procedure. For cutback asphalt mix designs using an average aggregate, tests should be scheduled using the estimated optimum asphalt content with alternate samples 0.5 percent below and 0.5 percent above (total of three samples, each with a different asphalt content).

In addition, swell test and moisture vapor susceptibility test specimens are prepared in duplicate at the "estimated optimum" asphalt content. Thus, for the normal mix design study a total of seven test specimens will usually be required. Although each test specimen will normally require only 1200 g of aggregate, the minimum aggregate requirements for one series of test specimens should be at least 18 kg (40 lb) to provide for check tests that may be required.

G.10 EQUIPMENT.—The equipment required for the preparation of test specimens is:
  (a) *Pans,* 250 millimetres (10 in.) diameter x 50 millimetres (2 in.) deep, for quartering and mixing fine aggregate.
  (b) *Pans,* 200 mm (8 in.) diameter x 45 mm (1 3/4 in.) deep, for batching and heating aggregates.
  (c) *Pans,* 305 mm (12 in.) diameter x 64 mm (2 1/2 in.) deep for mixing aggregate and asphalt.
  (d) *Pans,* 280 mm (11 in.) x 180 mm (7 in.) x 40 mm (1 1/2 in.) for curing mix.
  (e) *Sample Splitter,* large, for mixing and quartering fine aggregate.
  (f) *Hot Plate,* electric, at least 460 mm (18 in.) x 305 mm (12 in.), plate surface, for heating aggregates, asphalt and equipment as required.
  (g) *Oven,* large, thermostatically controlled, capable of 110° C (230° F) temperature.
  (h) *Oven,* large, thermostatically controlled, capable of 60° C (140° F) temperature.
  (i) *Oven,* large, for drying and preheating, capable of temperatures up to 163° C (325° F).
  (j) *Scoop,* large, for handling hot aggregates.
  (k) *Beakers,* 800 ml, for adding asphalt.
  (l) *Thermometer,* armored, 38° C (100° F) to 204° C (400° F).
  (m) *Balance,* 20-kg capacity, sensitive to 1.0 g, for weighing aggregates and asphalt.
  (n) *Mixing Trowel,* small pointed.
  (o) *Mixing Spoon,* large.
  (p) *Mechanical Mixer* (optional).
  (q) *Mechanical Compactor* designed to consolidate the material by a series of individual "kneading action" impressions made by a roving ram having a face shaped as a sector of a 101.6 mm (4 in.) diameter circle. The compactor must be capable of exerting a force of 34.5 kPa (500 psi) under the tamper foot. Accessories with the compactor should include two mold holders, a feeder trough 460 mm (18 in.) long x 102 mm (4 in.) wide x 64 mm (2 1/2 in.) deep and a paddle shaped to fit trough, and a bullet-nosed steel rod 9.5 mm (3/8 in.) diameter x 405 mm (16 in.) long.
  (r) *Compaction Molds,* steel, 101.6 mm (4 in.) inside diameter x 127 mm (5 in.) high x 6.4 mm (1/4 in.) wall thickness.
  (s) *Paper Disks,* heavy paper, 100 mm (3 15/16 in.) diameter, to place in bottom of mold during compaction.
  (t) *Compression Machine,* hydraulic, 222.4 kN (50,000 lb) capacity.
  (u) *Gloves,* heavy and sturdy, for handling hot equipment.

(Note: See Articles G.04 and G.15 for additional equipment requirements.)

G.11 BATCH WEIGHTS.—

(a) Compute batch weights for the blend and gradation of aggregates desired. A detailed procedure for computing batch weights is presented in MS-2. As a guide, the amount of mixture required for a specimen is:

Stabilometer test	1,200 to 1,400 g
Swell test	1,200 to 1,400 g
Moisture vapor susceptibility test	1,100 g
Moisture test	500 g

(b) The dry weight of the aggregate for the stabilometer specimens is that which will produce a compacted specimen of 63.5 ± 1.3 mm (2.5 ± 0.05 in.) in height. This will normally be about 1200 grams. To determine this exact batch weight it is generally desirable to prepare a trial specimen prior to preparing the aggregate batches. If the trial specimen height falls outside the limits, the amount of aggregate used for the specimen may be adjusted:

For International System of Units (SI),

$$\text{Adjusted mass of aggregate} = \frac{63.5 \times (\text{mass of aggregate used})}{\text{Specimen height obtained (mm)}}$$

For U.S. Customary Units,

$$\text{Adjusted weight of aggregate} = \frac{2.5 \times (\text{weight of aggregate used})}{\text{Specimen height obtained (in.)}}$$

G.12 PREPARATION OF BATCH MIXES.—

(a) Weigh the various-sized fractions of dry aggregates into suitable pans in accordance with accumulative batch weights.
(b) Cutback asphalt is added to unheated aggregate at the minimum temperature that will permit ready pouring. Do not heat the mixture more than necessary to obtain a uniform blend. Mechanical or hand mixing may be used. Mix aggregates and asphalt until all particles are coated.
(c) After mixing is complete, transfer the batch mix to a suitable flat pan and place in oven for a fifteen-hour curing period at 60° C ± 2.8° C (140° F ± 5° F). The oven used for curing asphalt should preferably be equipped for forced draft air circulation.

(Note: To prevent any possibility of ignition of hydrocarbon vapors, an explosion-proof motor should be used.)

G.13 COMPACTION.—The compaction of the test specimen is accomplished by means of the mechanical compactor that imports a kneading action consolidation by a series of individual impressions made with a ram having a face shaped as a sector of a 101.6 mm (4 in.) diameter circle. At each application of the ram a pressure of 3.45 MPa (500 psi) is applied, subjecting the specimen to a kneading action without impact over an area of approximately 2000 mm^2 (3.1 in.^{2}). Each pressure is maintained for approximately 0.4 of a second. The detailed compaction procedure is:

For Stabilometer Specimens:
(a) Preheat compaction molds to approximately the temperature to be used for the compaction of the mix.
(b) Heat compactor foot to a temperature that will prevent the mix from adhering to it. Temperature of the compactor foot may be controlled by a variable transformer.
(c) Place the compaction mold in position in the mold holder with a 101.6 mm (4 in.) diameter paper disc inserted to cover the base plate. In order to have the base plate act as a free-fitting plunger during the compaction operation, a steel shim 6.4 mm (1/4 in.) thick × 19.0 mm (3/4 in.) wide × 63.5 mm (2 1/2 in.) long is temporarily placed under the edge of the mold.
(d) Spread the prepared mixture (see batch mix preparation) uniformly in feeder trough. The trough should be preheated to approximately 60° C (140° F). With a paddle made to fit the shape of the trough, transfer approximately one-half of the mixture to the compaction mold (See Figure G-2).
(e) Rod the portion of the mix in the mold 20 times in the center of the mass and 20 times around the edge with the bullet-nosed steel rod (Figure G-3). Transfer the remainder of the sample to the mold and repeat the rodding procedure.
(f) Place mold assembly into position on the mechanical compactor and apply approximately 20 tamping blows at 1.7 MPa (250 psi) pressure to accomplish a semi-compacted condition of the mix so that it will not be unduly disturbed when the full load is applied. The exact number of blows to accomplish the semi-compaction shall be determined by observation. The number of blows may vary between 10 and 50, depending upon the type of material, and in some instances it may not be possible to accomplish the compaction in the mechanical compactor because of undue movement of the mixture under the compactor foot. In these instances use a 177.93 kN (40,000 lb.) static load applied by the double plunger method, in which a free-fitting plunger is placed below the sample as well as on top. Apply the load at the rate of 1.3 mm (0.05 in.) per minute and hold for 30 ± 5 seconds.

Figure G-2. Transfer of mix to mold   **Figure G-3. Rodding mix in mold**

(g) After the semi-compaction, remove shim and release mold tightening screw sufficiently to allow free up-and-down movement of mold.

(h) To complete compaction in the mechanical compactor, increase compactor foot pressure to 3.45 MPa (500 psi) and apply 150 tamping blows.

(i) The mold and specimen are placed in the 60° C (140° F) oven for 1 1/2 hours, after which a "leveling-off" load of 6.9 MPa (1,000 psi) is applied by the "double-plunger" method [head speed 0.02 mm per second (0.05 in/min)] and released immediately. (Note: The specimen shall not be pushed to the opposite end of the mold.)

For Swell Test Specimens:

(a) Prepare compaction mold by placing a paraffin-impregnated strip of ordinary wrapping paper 19 mm (3/4 in.) wide, around the inside of mold 13 mm (1/2 in.) to 19 mm (3/4 in.) from the bottom to prevent water from escaping from between the specimen and the mold during the water immersion period of the test. The paper strip is dipped in melted paraffin and applied while hot. Compaction molds are not preheated for swell test specimens.

(b) The remainder of the compaction procedure for swell test specimens is the same as for the stabilometer test specimens with this exception:

When compaction is completed in the mechanical compactor, remove mold and specimen from compactor, invert mold and push specimen to the opposite end of mold. Apply a 6.9 MPa (1,000 psi) static load [head speed 0.02 mm per second (0.05 in/min)] with the "original" top surface supported on the lower platen of testing press. It is advisable to place a piece of heavy paper under the specimen to prevent damage to this lower platen.

For Moisture Vapor Susceptibility Specimens:
(a) Compact moisture vapor susceptibility specimens the same as stabilometer specimens except use stainless steel molds.
(b) After compaction is completed in the mechanical compactor and the mold and specimens have been placed in the 60° C (140° F) oven for one hour, invert the mold and place on the moisture vapor susceptibility pressing standard. Press the specimen down through the mold and seat on the pressing standard. Apply a 56.0 kN (12,600 lb) static leveling load.

D. Test Procedures

G.14 GENERAL.—The compacted test specimens are used in these tests:
(a) Swell
(b) Stabilometer
(c) Moisture Vapor Susceptibility
(d) Cohesiometer

The swell test is performed only on specimens prepared for that purpose. The stabilometer test is performed on specimens prepared for stabilometer tests. After the stabilometer test is performed, the cohesiometer test is performed on the same specimens. Moisture vapor susceptibility test specimens are subjected to the stabilometer test, followed by the cohesiometer test, and finally the moisture content test.

G.15 EQUIPMENT.—The equipment required for the testing of the 102 mm (4 in.) diameter specimens is:
(a) *Bronze Disks,* perforated, 98.4 mm (3 7/8 in.) diameter x 3.2 mm (1/8 in.) thick, with adjustable stem, for swell measurement. (See Figure G-4.)
(b) *Dial Gauge,* mounted on tripod, with reading accuracy to 0.025 mm (0.001 in.) (See Figure G-4.)
(c) *Scale,* graduated to read the volumetric contents of a 101.6 mm (4 in.) inside-diameter mold at 25 ml intervals, for measuring percolation of water during swell test.
(d) *Pans,* aluminum, 190 mm (7 1/2 in.) diameter x 64 mm (2 1/2 in.) deep.

(e) *Hveem Stabilometer* (see Figures G-5 and G-6), complete with accessories including adjustable base, assembly tool, steel follower, and rubber bulb for introducing air into system.
(f) *Scale* or other measuring device to accurately determine height of compacted test specimen.
(g) *Cohesiometer* (see Figure G-7), complete with insulated heating cabinet and other accessories.
(h) *Aluminum seal cap,* 101.6 mm (4 in.) diameter, 18-20 gauge (See Figure G-8.)
(i) *Circular Felt Pad,* 102 mm (4 in.) diameter, 6 mm (1/4 in.) thick.
(j) *Felt Strip Wick,* 6 mm (1/4 in.) x 50 mm (2 in.) x 190 mm (7 1/2 in.)
(k) *Metal Spring Retaining Clamp.*
(l) *Tin-Plated Pan,* 32 mm (1 1/4 in.) high x 97 mm (3 13/16 in.) in diameter.
(m) *Special Pressing Standard* for applying aluminum seal caps (See Figure G-9).

G.16 SWELL TEST.—

(a) Allow compacted specimen for swell test to stand at room temperature for at least one hour (this is done to permit rebound after compaction.)
(b) Place mold and specimen in 190 mm (7 1/2 in.) diameter x 64 mm (2 1/2 in.) deep aluminum pan. (See Figure G-4.)
(c) Place perforated bronze disk on specimen, position tripod with dial gauge on mold, and set adjustable stem to give a reading of 2.54 mm (0.10 in.) on dial gauge. (See Figure G-4.)
(d) Introduce 500 ml of water to mold on top of specimen and measure distance from top of mold to surface of water with graduated scale.
(e) After 24 hours read dial gauge and record the change as *swell* to nearest 0.025 mm (0.001 in.) Also measure distance from top of mold to surface of water with graduated scale and record the change as *permeability* or the amount of water in milliliters that percolated into and/or through the test specimen.

G.17 STABILOMETER TEST.—(See Figures G-5 and G-6.)

(a) Place specimens for stabilometer tests (compacted and contained in mold) in oven at 60° C (140° F) for a minimum of one hour before testing.
(b) Adjust compression machine for a head speed of 0.02 mm per second (0.05 in./min) with no load applied.
(c) Check displacement of stabilometer with metal dummy specimen and, if necessary, adjust to give 2.00 ± 0.05 turns.
(d) Adjust the stabilometer base so that the distance from the bottom of the upper tapered ring to the top of the base is 89 mm (3.5 in.).
(e) Every effort should be made to fabricate test specimens with an overall height between 61 mm (2.40 in.) and 66 mm (2.60 in.); however, if the height is outside of this range the stabilometer value should be corrected as indicated in Figure G-10.

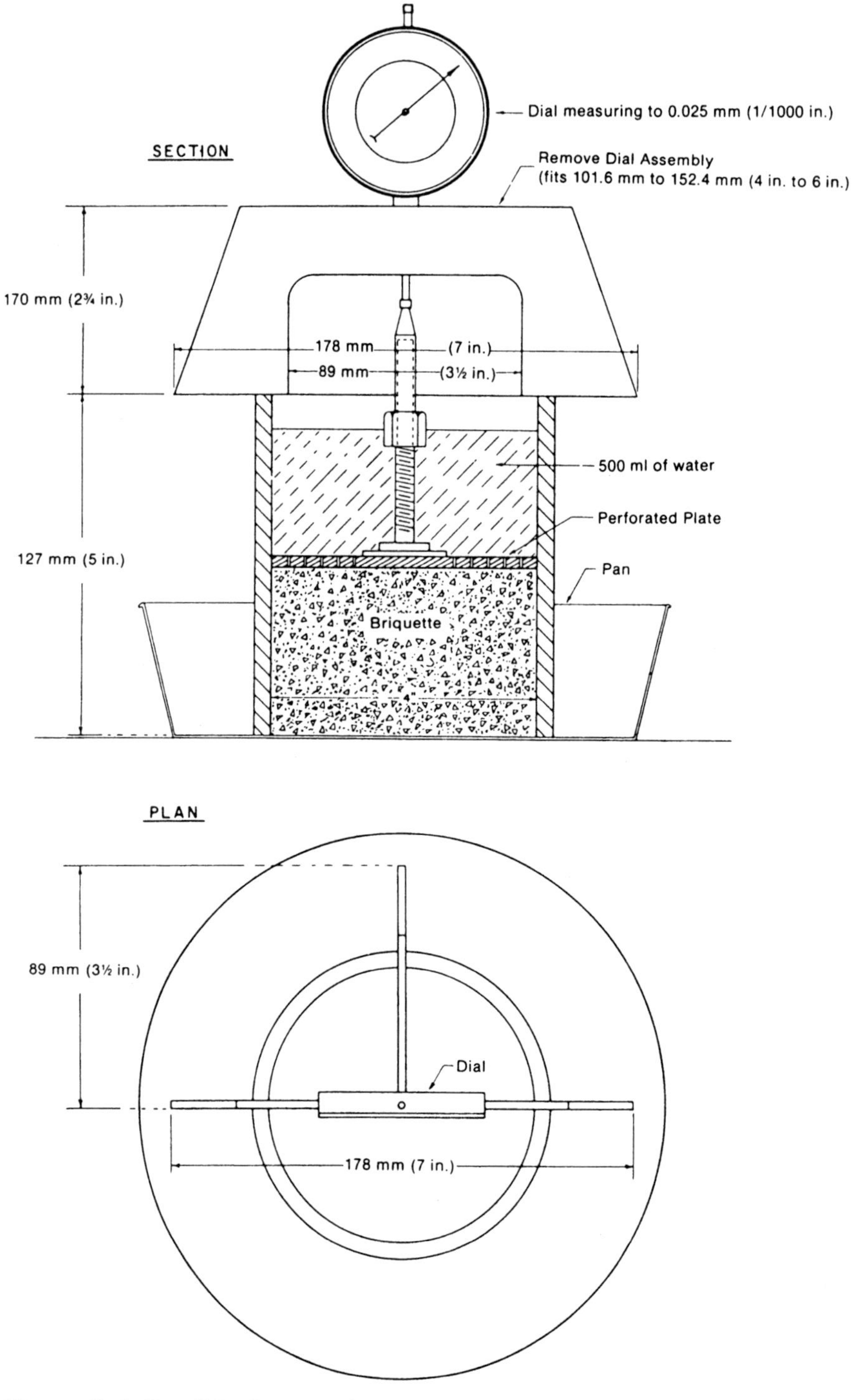

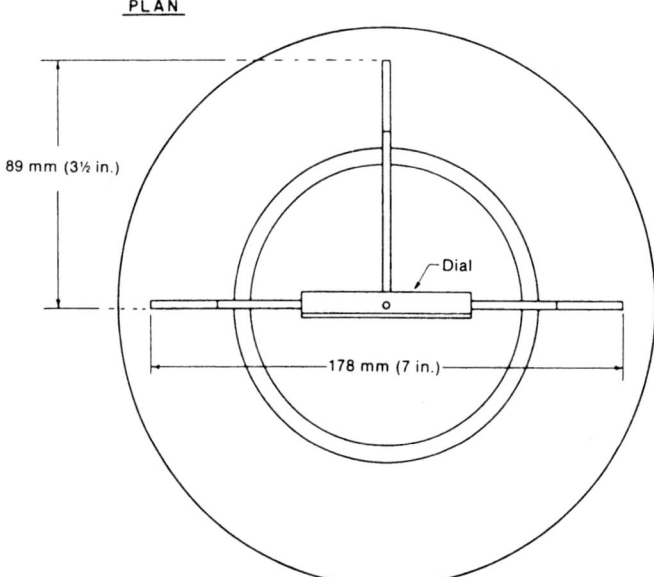

Figure G-4. Swell test apparatus

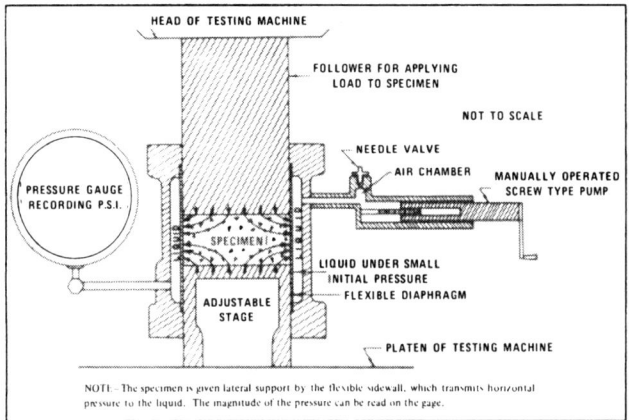

Figure G-5. Diagram showing principal features of Hveem stabilometer

Figure G-6. Hveem stabilometer

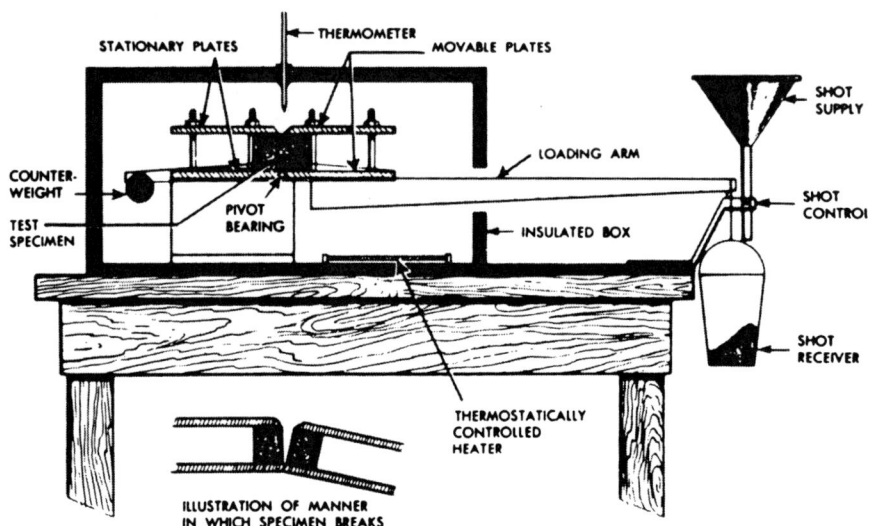

Figure G-7. Diagram showing features of Hveem Cohesiometer

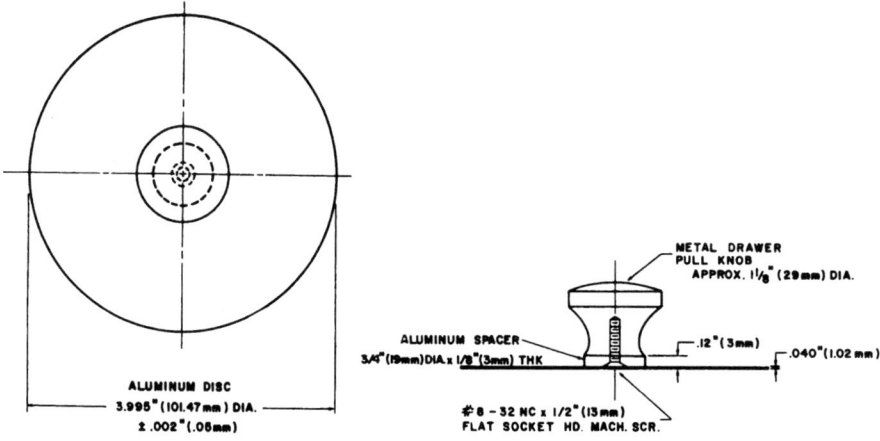

Figure G-8. Aluminum seal cap

MS-14   Appendix G 137

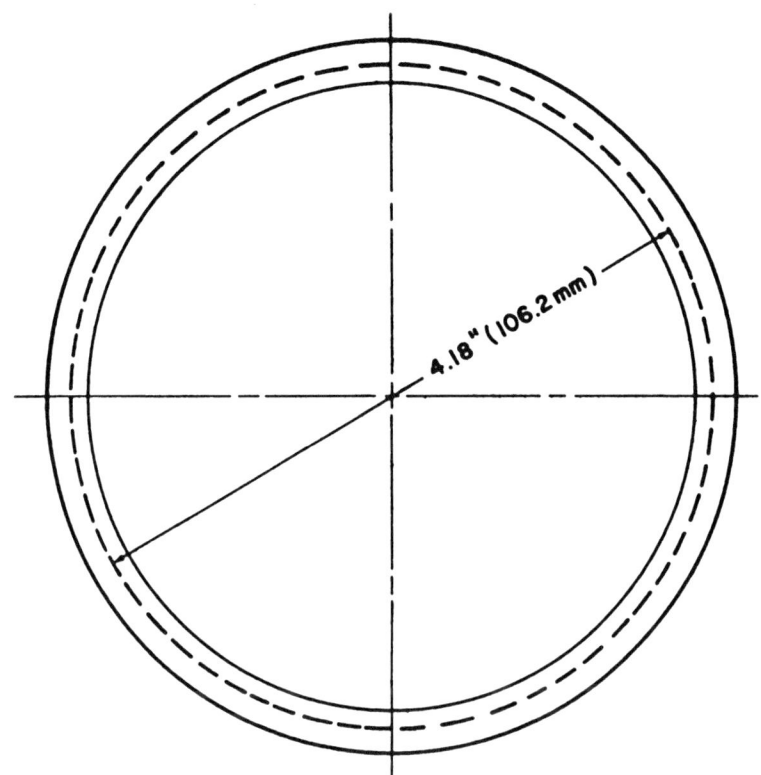

MILD STEEL

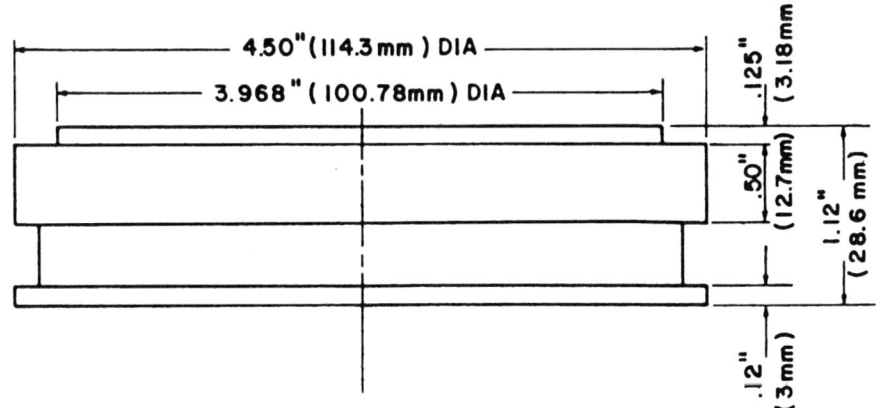

Figure G-9. Pressing standard

(f) Remove mold containing specimen from oven and place on top of stabilometer. Using the plunger, hand lever, and fulcrum, force the specimen from the mold into the stabilometer. Take care that the specimen goes in straight and is firmly seated.

(g) Place follower on top of specimen and position the entire assembly in compression machine for testing.

(h) Using a displacement pump, raise the pressure in the stabilometer system until test gauge reads exactly 34.5 kPa (5 psi). Then tap test gauge lightly to assure an accurate reading.

(i) Close displacement pump valve, taking care not to disturb the 34.5 kPa (5 psi) initial pressure. (This step is omitted on stabilometers that are not provided with the displacement pump valve.)

(j) Apply test loads with compression machine using a head speed of 0.02 mm per second (0.05 in/min). Record readings of test gauge at test loads of 2.22 kN (500 lb), 4.45 kN (1,000 lb) and each 4.45 kN (1,000 lb) thereafter up to a maximum of 26.69 kN (6,000 lb).

(k) Immediately after recording the reading under maximum load [26.69 kN (6,000 lb)], reduce total load on specimen to 4.45 kN (1,000 lb).

(l) Open the displacement pump angle valve and by means of the displacement pump, adjust test gauge to 34.5 kPa (5 psi) (this will result in a reduction in the applied press load which is normal and for which no compensation is made).

(m) Adjust dial gauge on pump to zero by means of small thumbscrew.

(n) Turn displacement pump handle rapidly (two turns per second) and smoothly to right (clockwise) until a pressure of 690 kPa (100 psi) is recorded on test gauge. (During this operation the load registered on the testing press will increase and in some cases exceed the initial 4.45 kN (1,000 lb) load. This change in load is normal and no adjustment or compensation is required under the present procedure). Record the exact number of turns required to increase the test gauge reading from 34.5 kPa (5 psi) to 690 kPa (100 psi) as *displacement* on specimen. [2.5 mm (0.1 in.) dial reading is equivalent to one turn displacement.]

(o) After recording the displacement, first remove the test load and reduce pressure on test gauge to zero, by means of displacement pump; then back off displacement pump an additional three turns and remove specimen from stabilometer chamber.

G.18 MOISTURE VAPOR SUSCEPTIBILITY TEST.—

(a) Place aluminum seal cap on the compacted surface and seal the edges to prevent the escape of moisture vapor. To seal, use a solution consisting of air-blown asphalt that has been dissolved in ethylene dichloride to produce a consistency comparable to that of ordinary paint.

(b) Place circular felt pad, which has previously been soaked in water, against the bottom surface of the test specimen. Place presoaked felt strip wick in contact with felt pad, the wick to be held in place with a metal spring clamp (See Figure G-11.)
(c) Insert pan of water up into the mold making certain that the free ends of the wick are immersed.
(d) Place assembly in a 60° C (140° F) oven for a continuous period of 75 hours.
(e) Remove the aluminum seal cap and wick assembly and obtain the stabilometer and cohesiometer values. The cohesiometer test should be performed immediately following the stabilometer test, to preserve the moisture absorbed during the 75-hour moisture vapor treatment.
(f) Determine percent of moisture absorbed by the specimen by subjecting 500 g of the mixture to ASTM Test Method D 1461.

Notes:
Water should be maintained in pan continuously during the 75 hours in the oven. Periodic visual checks may prevent having a dry pan at end of test.

G.19 COHESIOMETER TEST.—
(a) Cohesiometer tests are performed on the same specimens previously used in stabilometer and moisture vapor susceptibility tests.
(b) Place specimen in oven for approximately two hours at 60° C (140° F).
(c) Calibrate cohesiometer device so that the shot (or liquid) will flow into the receiving bucket at end of 0.76 m (30 in.) level arm at the rate of 1800 ± 20 g per minute.
(d) Adjust heating unit in cabinet of cohesiometer device to maintain a temperature of $60° C \pm 1.1° C$ ($140° F \pm 2° F$).
(e) Lock device in position with release pin. Remove specimen from oven and clamp firmly in position, centered, and with top plates parallel with top surface of specimen. Permit temperature in cohesiometer cabinet to attain $60° C \pm 1.1° C$ ($140° F \pm 2° F$) before starting the test.
(f) Pull release pin and allow shot flow to continue until the specimen breaks, indicated by a sudden drop of beam.
(g) In the event that the specimen is flexible or ductile rather than brittle, the flow of shot is stopped when the end of the 0.76 m (30 in.) beam has lowered 13 mm (1/2 in.) from horizontal.
(h) Weigh shot caught in receiving bucket to the nearest gram and record as shot weight.

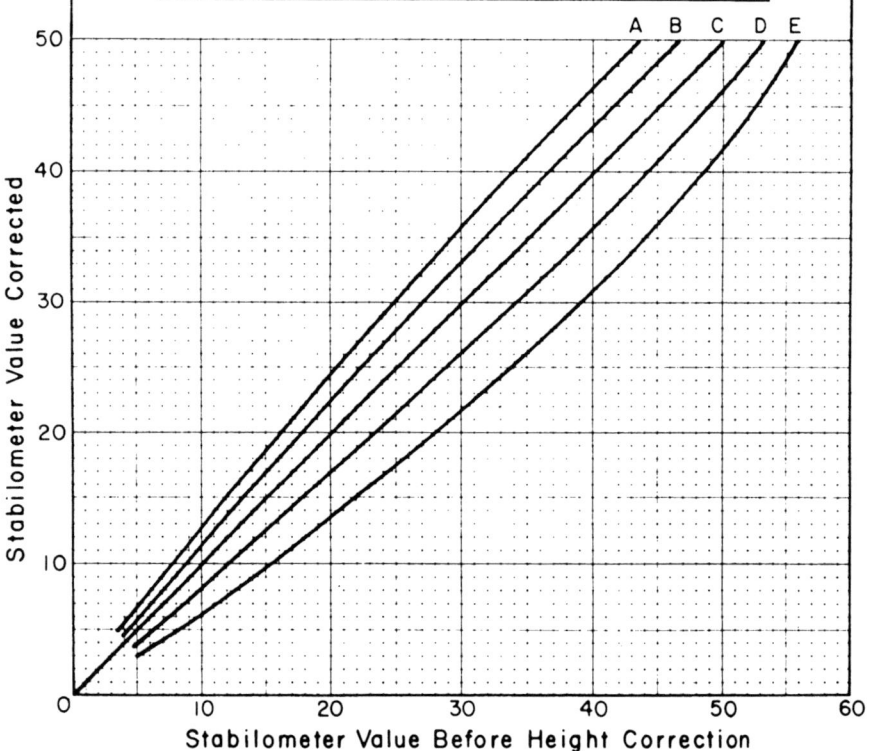

Figure G-10. Chart for correcting stabilometer value to effective specimen height of 64 mm (2.5 in.)

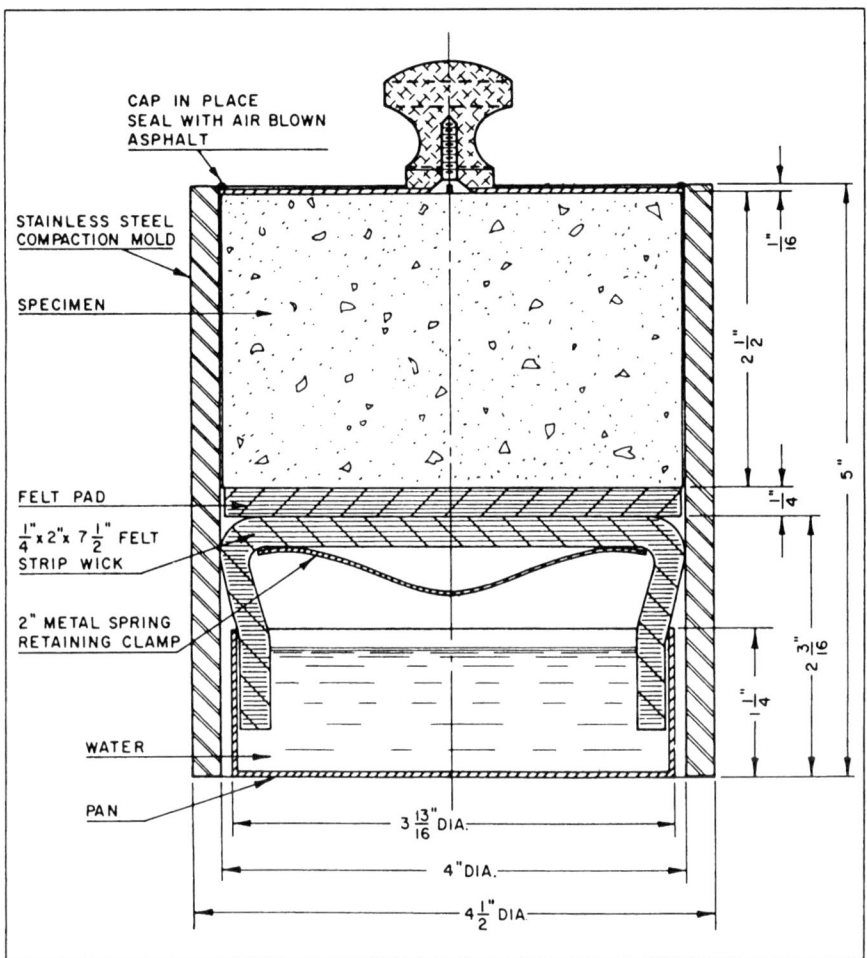

Figure G-11. Moisture vapor susceptibility test. (To convert inches to millimeters multiply by 25.4)

E. Interpretation of Test Data

G.20 CALCULATIONS.—There are no calculations required for the swell test since the results are reported directly as differences. The remainder of the calculations are:

(a) *Stabilometer Value.* Calculate as:

$$S = \frac{22.2}{\left(\dfrac{P_h \times D_2}{P_v - P_h}\right) + 0.222}$$

where S = stabilometer value
D_2 = displacement on specimen
P_v = vertical pressure [typically 2,758 kPa (400 psi) = 22.24 kN (5,000 lb) total load.]
P_h = horizontal pressure = stabilometer pressure gauge reading taken at the instant P_v is 2,758 kPa (400 psi) or 22.24 kN (5,000 lb) total load.

(b) *Cohesiometer Value.* Calculate as:

$$C = \frac{L}{W(0.20H + 0.044 H^2)}$$

where C = cohesiometer value (g per 1 in. width corrected to a 3 in. height)
L = weight of shot in grams
W = diameter or width of specimen in inches
H = height of specimen in inches

Note: This equation may only be used with U.S. Customary units. Metric units must be converted to obtain the correct cohesiometer value.

Cohesiometer values may also be obtained by multiplying the weight of shot necessary to break the specimen by factors established for various heights of 101.6 mm (4 in.) diameter (or width) specimens. The factors used are:

Height			Height		
mm	(in.)	Factor	mm	(in.)	Factor
55.9	(2.20)	.383	63.5	(2.50)	.323
57.2	(2.25)	.372	64.8	(2.55)	.314
58.4	(2.30)	.361	66.0	(2.60)	.306
59.7	(2.35)	.351	67.3	(2.65)	.298
61.0	(2.40)	.341	68.6	(2.70)	.290
62.2	(2.45)	.332	69.8	(2.75)	.283

Example: Assume that it takes 600 g of shot to break a certain specimen which has a 101.6 mm (4 in.) diameter and 63.5 mm (2.50 in.) height. Cohesiometer value = 600 x .323 = 194.

(c) *Moisture Vapor Susceptibility*. Report results as numerical values obtained in the stabilometer and cohesiometer tests and as percent of moisture obtained in the moisture test.

G.21 DESIGN CRITERIA.—Suitability of cutback asphalt mixtures is based on (1) stability, (2) volume change or swell when exposed to water, and (3) stability change when exposed to moisture vapor. Suggested criteria for cutback asphalt mixtures are given in Table G-1.

In evaluating stabilometer and cohesiometer test results, a low value in one may be somewhat compensated by a high value in the other. Visual inspection should always be part of the test results for determining a recommended optimum asphalt content.

In special cases where asphalt content determined from the C.K.E. test is not consistent with test results because of unusual characteristics of aggregates, it is necessary to select a recommended optimum asphalt content by means of stabilometer values and visual inspection.

Table G-1. Suggested Criteria for Cutback Asphalt Mixes

Test	Requirements
Stabilometer value	30 min.
Moisture Vapor Susceptibility (Stabilometer value)	20 min.
Swell	0.76 mm max. (0.030 in.) max.

Appendix H. **Marshall Method for Cutback Asphalt-Aggregate Cold Mixture Design**

A. General

H.01 APPLICATION.—This design method for cutback asphalt paving mixtures is based on the Marshall Method of Mix Design as described in *Mix Design Methods for Asphalt Concrete and Other Hot-Mix Types,* MS-2, Asphalt Institute. The method and recommended test criteria are applicable to paving and maintenance mixtures containing any grade of the medium curing (MC) and slow curing (SC) cutback asphalts and well-graded mineral aggregates with maximum sizes of 25 mm (1 in.) or less. This includes road mixes prepared at ambient temperatures as well as plant mixes that require mixing at elevated temperatures. This method is used for laboratory design only. It is not applicable to field control of cutback asphalt mixtures.

H.02 OUTLINE OF METHOD.—The procedure for the Marshall method starts with the preparation of test specimens. Preliminary to this operation it is required that:
 (a) the materials proposed for use meet the requirements of the project specifications;
 (b) aggregate blend combinations meet the gradation requirements of the project specifications; and
 (c) for use in density and voids analyses, the bulk specific gravity of all aggregates used in the blend, and the specific gravity of the asphalt cement, are determined.

These requirements are matters of routine testing, specifications and laboratory technique that must be considered but that are not unique to any particular mix design method. (Refer to MS-2 for additional information on the preparation and analysis of aggregates.)

The Marshall method uses standard test specimens of 64 mm (2 1/2 in.) height and 102 mm (4 in.) diameter. These are prepared using a specified procedure for heating, mixing and compacting the asphalt-aggregate mixtures. The two principal features of the Marshall method of mix design are a density-voids analysis and a stability-flow test of the compacted test specimens.

The stability of the test specimen is the maximum load resistance in Newtons (lb) that the standard test specimen will develop at 25° C (77° F) when tested as outlined hereinafter. The flow value is the total movement or deformation, in units of 0.25 mm (0.01 in.) occurring in the specimen between no load and maximum load during the stability test.

The design procedure consists of these steps:
 (a) An aggregate meeting the requirements of the project specifications is selected.

(b) A type and grade of cutback asphalt is selected according to the type of aggregate, mixing and placing equipment, and climatic conditions.
(c) Test mixes with various binder contents are prepared and compacted as in the case of asphalt concrete, except that the compaction temperature is based on the viscosity of the asphalt after it has been cured to a specified percentage of solvent loss.
(d) The test specimens are analyzed for voids content and voids in mineral aggregate and the Marshall stability and flow are determined at a temperature of 25° C (77° F).
(e) The effect of water on stability and flow is determined by subjecting a set of test specimens to a four-day immersion period in a water bath maintained at 25° C (77° F).
(f) The optimum asphalt content is chosen as the percentage of cutback asphalt at which the paving mixture, after it has been cured to the specified solvent loss, best satisfies all the design criteria.
(g) Design criteria form the basis for determining if the paving mixture will be satisfactory at the optimum asphalt content.

B. Preparation of Test Specimens

H.03 GENERAL.—In determining the optimum asphalt content for a particular aggregate or combination of aggregates, a series of test specimens is prepared with a range of asphalt contents, so that the test data show the influence of asphalt content on the mix properties. Tests should be scheduled on the basis of no more than 1 percent and preferably 0.5 percent increments of asphalt content with at least two asphalt contents above optimum and at least two asphalt contents below optimum. To establish the asphalt contents for use in these laboratory tests, the optimum asphalt content must *first* be estimated.

To provide adequate data, triplicate test specimens are usually prepared for each asphalt content used. Thus, a design study using six different asphalt contents will normally require at least eighteen test specimens. An additional six test specimens are required for the evaluation of the effect of water on stability and flow. Each test specimen will usually require approximately 1200 g of aggregate. Therefore, the minimum aggregate requirements for the preparation of test specimens with a given composition will be approximately 29 kg (65 lb). An extra quantity of aggregate is also required for sieve analysis and specific gravity determinations. Four litres (one gallon) of cutback asphalt will usually be adequate.

H.04 EQUIPMENT.—The equipment required for the preparation of test specimens is:
(a) *Pans,* metal, flat bottom, for heating aggregates.
(b) *Pans,* metal, round, approximately 4-litres (4-qt.) capacity, for mixing asphalt and aggregate.
(c) *Oven and Hot Plate,* electric, for heating aggregates, asphalt, and equipment as required.

(d) *Scoop,* for batching aggregates.
(e) *Containers,* gill-type tins, beakers, pouring pots, or sauce pans, for heating asphalt.
(f) *Thermometers,* armored, glass, or dial-type with metal stem, 10° C (50° F) to 232° C (450° F), for determining temperature of aggregates, asphalt and asphalt mixtures.
(g) *Balance,* 5-kg capacity, sensitive to 1 gm, for weighing aggregates and asphalt. *Balance,* 2-kg capacity, sensitive to 0.1 gm, for weighing compacted specimens.
(h) *Mixing Spoon,* large, or *Trowel,* small.
(i) *Spatula,* large.
(j) *Mechanical Mixer* (optional), commercial bread dough mixer 4-litre (4-qt.) capacity or larger, equipped with two metal mixing bowls and two wire stirrers.
(k) *Boiling Water Bath,* consisting of hot plate and bucket for water, for heating compaction hammer and mold.
(l) *Compaction Pedestal,* consisting of a 200 x 200 x 460 mm (8 x 8 x 18 in.) wooden post capped with a 305 x 305 x 25 mm (12 x 12 x 1 in.) steel plate. The wooden post should be oak, yellow pine or other wood having a dry weight of 673 to 769 kg/m^3 (42 to 48 lb/ft^3). The wooden post should be secured by four angle brackets to a solid concrete slab. The steel cap should be firmly fastened to the post. The pedestal should be installed so that the post is plumb, the cap level, and the entire assembly free from movement during compaction.
*(m) *Compaction Mold,* consisting of a base plate, forming mold, and collar extension. The forming mold has an inside diameter of 101.6 mm (4 in.) and a height of approximately 75 mm (3 in.); the base plate and collar extension are designed to be interchangeable with either end of the forming mold.
*(n) *Compaction Hammer,* consisting of a flat circular tamping face 98.4 mm (3 7/8 in.) diameter and equipped with a 4.5 kg (10 lb.) weight constructed to obtain a specified 457 mm (18 in.) height of drop.
*(o) *Mold Holder,* consisting of spring tension device designed to hold compaction mold in place on compaction pedestal.
(p) *Extrusion Jack* or *Arbor Press,* for extruding compacted specimens from mold.
(q) *Gloves,* welders, for handling hot equipment. *Gloves,* rubber, for removing specimens from water bath.
(r) *Marking Crayons,* for identifying test specimens.
(s) *Oven,* forced draft for curing mixes.
(Note: See additional equipment requirements in Article H.07.)

H.05 PREPARATION OF TEST SPECIMENS.—
(a) *Number of Specimens*—Prepare at least three, preferably five, specimens for each combination of aggregates and asphalt content.

*Marshall test apparatus shall conform to requirements of ASTM Test Method D1559.

(b) *Testing and Preparation of Aggregates*—Determine the gradation of each aggregate proposed for use by washing and sieving according to ASTM Test Methods C 117 and C 136. Also determine the bulk and apparent specific gravities of each aggregate according to ASTM Test Methods C 127 and C 128. Calculate and select an aggregate combination whose gradation deviates sufficiently from the maximum density curve for the specific maximum size to indicate that the minimum VMA requirement will be met. Calculate the bulk and apparent specific gravities of the combination of mineral aggregate selected. Dry aggregates to constant weight at 105° C (221° F) to 110° C (230° F) and separate by dry sieving into the desired size fractions. The fractions recommended as generally adequate are: 25 to 19 mm (1 to 3/4 in.), 19 to 12.7 mm (3/4 to 1/2 in.), 12.7 to 9.5 mm (1/2 to 3/8 in.), 9.5 to 4.75 mm (3/8 in. to No. 4), 4.75 to 2.36 mm (No. 4 to No. 8) and passing 2.36 mm (No. 8).

(c) *Determination of Mixing and Compaction Temperature*—The temperature to which the cutback asphalt must be heated to produce a viscosity of 170 ± 20 centistokes shall be established as the mixing temperature. The mixing temperature is determined by referring to a viscosity-temperature chart for the particular type and grade of asphalt being used. An example of such a chart is shown in Figure H-1. Such a chart may be obtained from the asphalt supplier.

Referring to a composition chart for the type of cutback asphalt being used (an example is given in Figure H-2), determine from its viscosity at 60° C (140° F) the percent solvent by weight. From the same composition chart also determine the viscosity at 60° C (140° F) of the same asphalt after it has lost 50 percent of its solvent. (Note: For certain specific conditions, other percentages of solvent loss may be used.) For maintenance mixes to be stockpiled, a solvent loss of 25 percent is suggested. From the viscosity-temperature graph of cutback asphalts made from the same base stocks (similar to the example in Figure H-1) determine the temperature at which the asphalt would have a viscosity of 280 ± 30 centistokes after having lost the desired percentage of its solvent. This will be the compaction temperature.

(d) *Preparation of Mold and Hammer*—Thoroughly clean the specimen mold assembly and the face of the compaction hammer and heat them in a boiling water bath or on the hot plate to a temperature between 93° C (200° F) and 149° C (300° F). Place a piece of waxed paper cut to size in the bottom of the mold before the mixture is placed in the mold.

(e) *Preparation of Mixtures*—Weigh into separate pans for each test specimen the amount of each size fraction required to produce a batch that will result in a compacted specimen 63.5 ± 1.3 mm (2.5 ± 0.1 in.) in height. This will normally be about 1200 g. It is generally desirable to prepare a trial specimen prior to preparing the aggregate batches. If the trial specimen

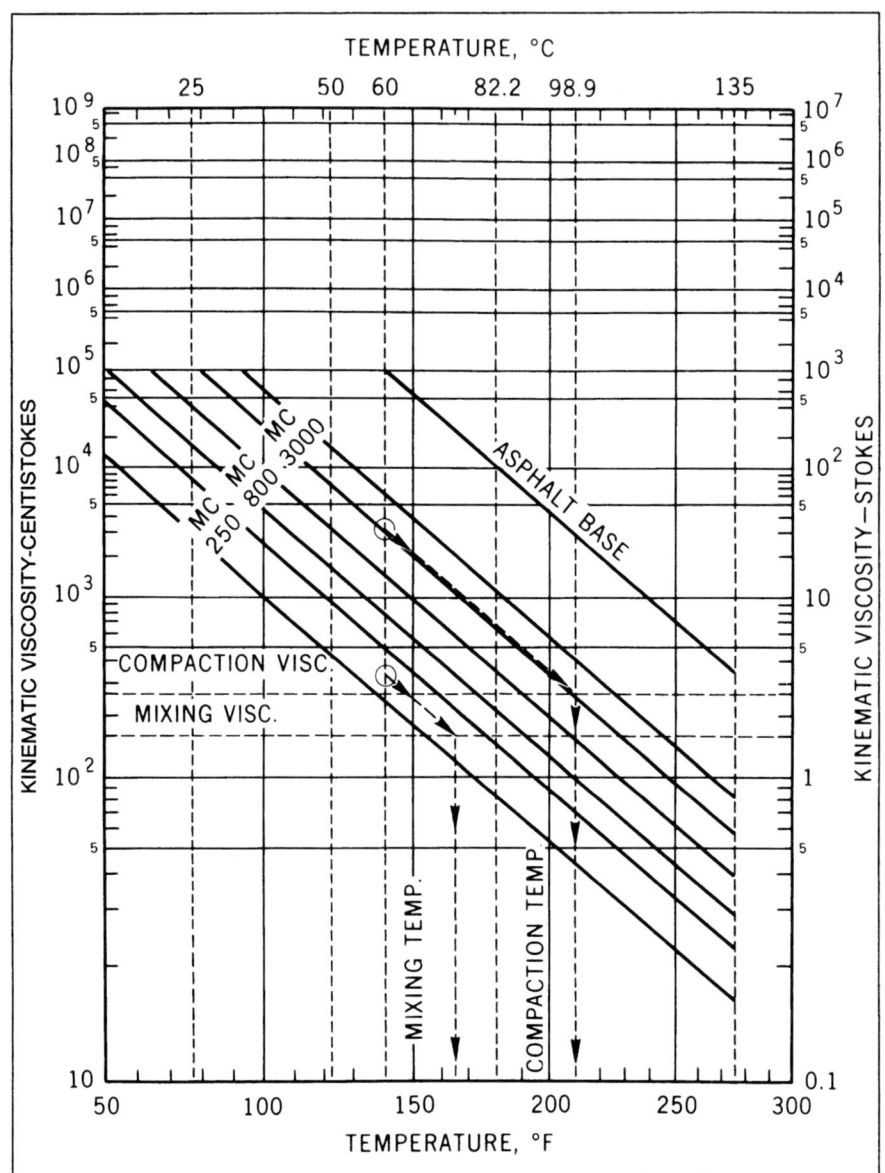

Figure H-1. Example of viscosity-temperature relationship for medium-curing cutback asphalt made from same base stocks.

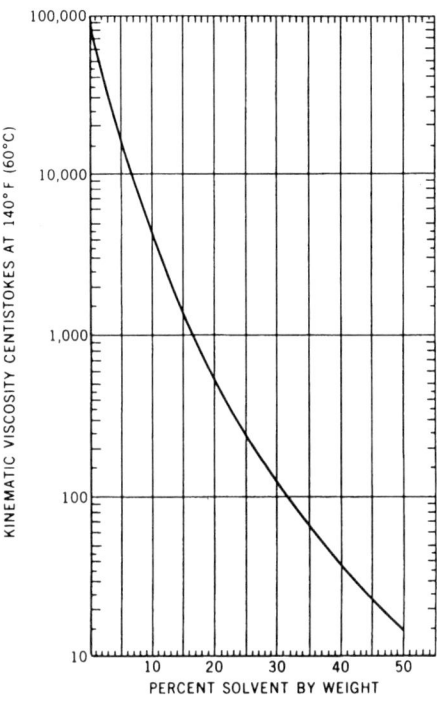

Figure H-2. Example of composition chart: medium-curing cutback asphalt made from the same base stocks.

height falls outside the limits, the amount of aggregate used for the specimen may be adjusted:

For International System of Units (SI),

$$\text{Adjusted weight of aggregate} = \frac{63.5 \times (\text{weight of aggregate used})}{\text{Specimen height obtained (mm)}}$$

For U.S. Customary Units,

$$\text{Adjusted weight of aggregate} = \frac{2.5 \times (\text{weight of aggregate used})}{\text{Specimen height obtained (in.)}}$$

Place the mixing pans in the oven and heat to a temperature approximately 14° C (25° F) above the mixing temperature. Heat the cutback asphalt, rapidly and carefully to avoid appreciable loss of solvent, to a temperature high enough to make it flow easily but not higher than the predetermined mixing temperature. Form a crater in the hot, blended aggregate, place the mixing pan and contents on a balance and weigh in the required amount of asphalt. At this point, the temperature of the aggregate and bituminous material shall be within the limits of the mixing temperature established in (c) above. Immediately introduce the mixing trowel in the mixing pan and determine the total weight of the mix components plus mixing equipment within 0.2 g. Mix the aggregate and cutback asphalt with the trowel until a uniform mixture is obtained. Care must be exercised not to lose any of the material during mixing and subsequent handling. A mechanical mixer may also be used, provided the mixing blade can be weighed with the mixing bowl and contents and introduced whole into the oven for curing without having to transfer the mix to another container and cause loss of material.

The mixtures are cured in a ventilated oven maintained at the compaction temperature plus approximately 11° C (20° F) to allow for heat loss during subsequent handling of the mix. Curing is carried out in the mixing bowl and controlled by verifying the weight at intervals of 15 minutes initially and less than 10 minutes as the weight of the mix at the predetermined solvent loss is approached. The mix may be turned over in the mixing pan during curing to accelerate solvent loss but care must be taken so as not to lose any of the material. All weights must be determined to ± 0.2 g.

(f) *Compaction of Specimens*—Place the entire batch in the mold, spade the mixture vigorously with a heated spatula or trowel 15 times around the perimeter and ten times over the interior. Remove the collar and smooth the surface to a slightly rounded shape. Temperature of the mixture immediately prior to compaction shall be within the limits of the compaction temperature established in (c); otherwise, it shall be discarded.

Replace the collar and place the mold assembly on the compaction pedestal in the mold holder. Apply 75 blows with the compaction hammer. Remove the base plate and collar, reverse and reassemble the mold. Apply the same number of compaction blows to the face of the reversed specimen. After compaction, remove the base plate and allow the specimen to cool to ambient temperature in its mold. Remove the specimen from the mold by means of an extrusion jack or other compression device, then place on a smooth, level surface until ready for testing. Any specimen varying by more than ± 2.5 mm (0.1 in.) from the specified thickness of 63 mm (2.5 in.) shall be rejected. Testing should not be started sooner than 16 hours after compaction. If the specimens are to be stored for more than 24 hours before testing, they shall be protected from exposure to the air by sealing them in close-fitting, airtight containers.

C. Test Procedure

H.06 GENERAL.—To complete the mix design, these tests and analyses are made:
- (a) *Uncompacted mixture.*
    - (1) Effective Specific Gravity
    - (2) Maximum Specific Gravity
- (b) *Compacted test specimens.*
    - (1) Bulk Specific Gravity
    - (2) Stability and Flow at 25° C (77° F)
    - (3) Stability retention after immersion in water at 25° C (77° F)
    - (4) Density and Voids Analysis

H.07 EQUIPMENT.—The equipment required for the testing of the 102 mm (4 in.) diameter x 64 mm (2 1/2 in.) height specimens is:

(a) *Marshall Testing Machine,* a compression testing device, conforming to ASTM Test Method D1559. It is designed to apply loads to test specimens through semi-circular testing heads at a constant rate of strain of 50.8 mm (2 in.) per minute. It is equipped with a calibrated proving ring for determining the applied testing load, a Marshall stability testing head for use in testing the specimen, and a Marshall flow meter for determining the amount of strain at the maximum load for test. A universal testing machine equipped with suitable load and deformation indicating devices may be used instead of the Marshall testing frame.

(b) *Air Bath.* The air bath shall be equipped with either a manual or an automatic control capable of maintaining the temperature at 25° C ± 0.6° C (77° F ± 1.0° F).

(c) *Water Bath,* at least 150 mm (6 in.) deep and thermostatically controlled at 25° C ± 1.0° C (77° F ± 1.8° F).

H.08 EFFECTIVE SPECIFIC GRAVITY DETERMINATION.—Combine prepared fractions of the mineral aggregate in the proper proportions to obtain the selected gradation and in sufficient quantity to produce a mixture of the minimum size specified by ASTM D 2041, "Maximum Specific Gravity of Bituminous Paving Mixtures." Using an asphalt cement, prepare a hot mix as described in Art. 3.05, MS-2, using a percentage of binder high enough to thoroughly coat the aggregate. Cool the loose mix and determine its maximum specific gravity according to ASTM D 2041. Repeat the same mix so as to have duplicate test results for maximum specific gravity. Calculate the effective specific gravity of the aggregate from the average maximum specific gravity, the percentage of asphalt cement and its specific gravity at 25° C (77° F), and the percentage of mineral aggregate according to the method given in Chapter VI, MS-2. In this procedure, it is assumed that the aggregate will absorb the same percentage of cutback asphalt as it does asphalt cement.

H.09 MAXIMUM SPECIFIC GRAVITY CALCULATION.—The maximum specific gravity is calculated for each binder content used in the mixes. The binder content is defined as the percentage of cured asphalt remaining after evaporation of the specified portion of solvent in the curing operation. Therefore, these data are used for the calculation: effective specific gravity of the aggregate, percentage of aggregate in the cured mix, percentage of binder in the cured mix, and specific gravity of the binder. In order to determine the specific gravity of the binder, a chart of specific gravity versus solvent content for the cutback asphalt being used, similar to that shown in the example (Figure H-3) is required.

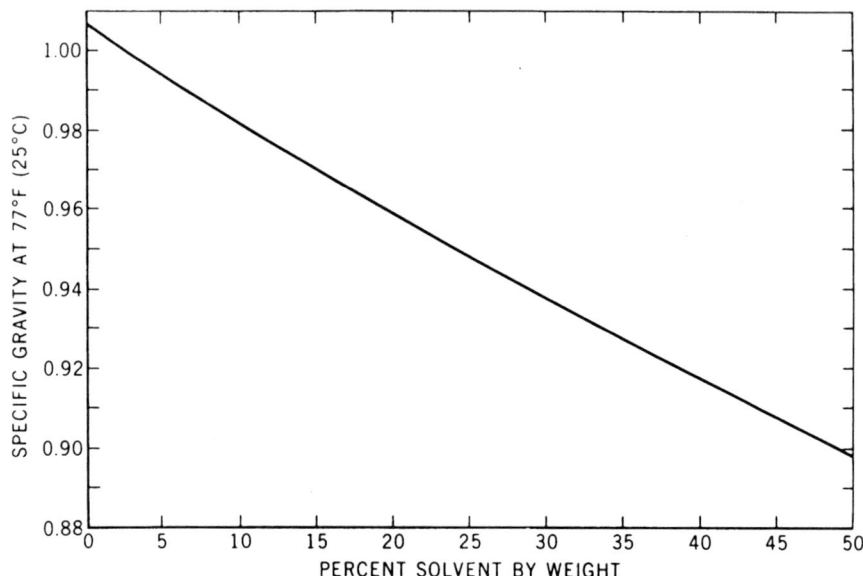

Figure H-3. Example of specific gravity at 25° C (77° F) of medium-curing cutback asphalts made from the same base stocks.

H.10 BULK SPECIFIC GRAVITY DETERMINATION.—Depending on the porosity and surface texture characteristics of the compacted test specimens, use either: ASTM D 1188, "Bulk Specific Gravity of Compacted Bituminous Mixtures Using Paraffin Coated Specimens," or ASTM D 2726, "Test for Bulk Specific Gravity of Compacted Bituminous Mixtures Using Saturated Surface Dry Specimens."

H.11 STABILITY AND FLOW TESTS.—After determining the bulk specific gravity, perform the stability and flow tests:
  (a) Condition the test specimens in the air bath at 25° C ± 0.7° C (77° F ± 1.0° F) for two hours before testing.
  (b) Thoroughly clean the guide rods and the inside surfaces of the test heads prior to making the test, and lubricate the guide rods so that the upper test head slides freely over them. The testing head temperature shall be maintained between 21.1 and 26.7° C (70 and 80° F) using a water bath when required. Check the flow meter and load measuring device for "zero" adjustment.
  (c) Remove a test specimen from air bath, place it in the lower testing head, then fit upper testing head into position and center complete assembly in the loading device. Place flow meter over marked guide rod.
  (d) Apply a testing load to specimen at constant rate of deformation of 50.8 mm (2 in.) per minute until failure occurs. The point of failure is defined by the maximum load reading obtained. The total number of Newtons (lb) required

to produce failure of the specimen at 25° C (77° F) shall be recorded as its Marshall stability value.

(e) While the stability test is in progress, hold the flow meter firmly in position over the guide rod and remove it the instant the maximum load starts to decrease. Note and record the indicated flow value in units of 0.25 mm (0.01 in.).

(f) Average the stability and flow values for all specimens with a given binder content. Values that are obviously in error shall not be included in the average.

(g) Prepare separate graphical plots showing stability at 25° C (77° F) and flow value vs. original cutback asphalt content.

H.12 DENSITY AND VOIDS ANALYSIS.—A density and voids analysis is made for each set of test specimens:

(a) Average the bulk specific gravity values for all test specimens with a given binder content; values obviously in error shall not be included in the average.

(b) Determine average unit weight in kg/m^3 for each binder content by multiplying the average bulk specific gravity value by 1000 (for unit weight in lb/ft^3, multiply by 62.4).

(c) Prepare a graphical plot of unit weight vs. original cutback asphalt content and draw a smooth curve which gives the best fit for all values.

(d) Read unit weight values directly from the plotted curve for each original cutback asphalt content tested and compute equivalent bulk specific gravity values by dividing kg/m^3 by 1000 (lb/ft^3 by 62.4). The values of bulk specific gravity thus obtained shall be used in further calculations of voids data.

(e) Using the maximum specific gravity corresponding to each binder content and the bulk specific gravity, calculate the percentage of air voids in each set of test specimens. Prepare a graphical plot of percent air voids vs. original cutback asphalt content and connect plotted points with a smooth curve.

(f) Using the bulk specific gravity of each series of test specimens, the binder content, and the bulk specific gravity of the aggregate, calculate the percentage of voids in the mineral aggregate (VMA). Plot VMA values vs. original cutback asphalt content and connect points with a smooth curve.

D. Interpretation of Test Data

H.13 PREPARATION OF DATA.—

(a) Average the flow values and stability values for all specimens of a given asphalt content. Values that are obviously in error shall not be included in the average.

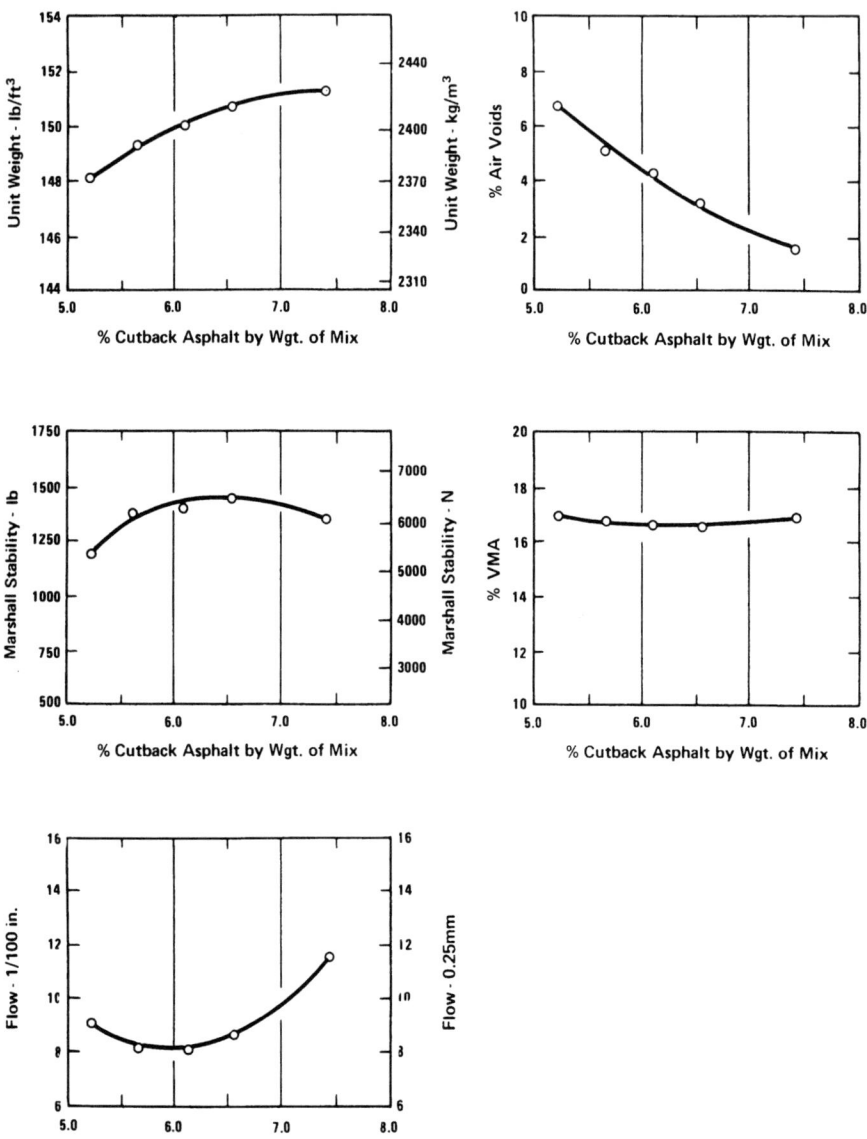

Figure H-4. Test property curves for cutback asphalt mixture design data by the Marshall Method (see also Table H-4).

(b) Prepare a separate graphical plot for these values (as illustrated in Figure H-4):
   (1) Stability vs. Asphalt Content
   (2) Flow vs. Asphalt Content
   (3) Unit Weight of Total Mix vs. Asphalt Content
   (4) Percent Air Voids vs. Asphalt Content
   (5) Percent Voids in Mineral Aggregate (VMA) vs. Asphalt Content
   In each graphical plot connect the plotted values with a smooth curve that obtains the "best-fit" for all values.

H.14 TERMS AND RELATIONS OF TEST DATA.—Because mineral aggregate normally absorbs less asphalt than water, the effective specific gravity is an intermediate value between the bulk and the apparent specific gravities. The three values should, therefore, be compared to verify the test data.

The test property curves, plotted as described above, have been found to follow a reasonably consistent pattern for dense-graded asphalt paving mixes. Trends generally noted are:
(a) The stability value increases with increasing asphalt content up to a maximum, after which the stability decreases.
(b) The flow value decreases to a minimum value then increases with increasing asphalt content. In many cases, the minimum value will occur at a lower asphalt content than the minimum used in the test specimens.
(c) The curve for unit weight of total mix is similar to the stability curve, except that the maximum unit weight normally (but not always) occurs at a slightly higher asphalt content than the maximum stability.
(d) The percent of air voids decreases with increasing asphalt content, ultimately approaching a minimum void content.
(e) The percent voids in the mineral aggregate (VMA) generally decreases to a minimum value then increases with increasing asphalt content.

H.15 DETERMINATION OF OPTIMUM ASPHALT CONTENT.—Select as the optimum asphalt content the percentage that will allow the maximum variation above and below it while all the design criteria are met. Because the percentage of air voids usually is the limiting characteristic, the optimum asphalt content is often selected as that corresponding to the median criteria (4 percent).

H.16 EFFECT OF WATER ON TEST RESULTS.—
(a) Prepare a set of at least six test specimens, containing the optimum percentage of cutback asphalt according to the procedure in Art. H.05.
(b) Determine the bulk specific gravity of each test specimen and compute the average. Separate the specimens into two groups so that the average bulk specific gravity of each group will be as close as possible to the overall average.
(c) Keep one group of test specimens in air at room temperature for 16 to 24 hours, then condition for two hours in the air bath at 25° C ± 0.6° C (77° F ± 1.0° F) before testing for stability and flow at 25° C (77° F).

(d) Place the other group of test specimens on transfer plates and keep in air at room temperature for 16 to 24 hours, then immerse for four days in a water bath maintained at 25° C ± 0.6° C (77° F ± 1.0° F) before testing for stability and flow at 25° C (77° F).

(e) Compute the percentage of retained stability after water immersion from the average Marshall stability of each group of test specimens.

H.17 CRITERIA FOR SATISFACTORY MIX.—The design criteria recommended for determining the suitability of designs for paving mixtures containing cutback asphalt are shown in Table H-1.

Table H-1. Marshall Design Criteria for Paving Mixtures Containing Cutback Asphalt

Test Property	Minimum	Maximum
Degree of Curing		
Percent solvent evaporated		
Maintenance Mixtures		25
Paving Mixtures		50
Number of Hammer Blows		
Hand Compactor		75
Percent Air Voids,		
Compacted Mix	3	5
Percent Voids in	See Figure H-5	
Mineral Aggregate (VMA)		
Stability, (N) lb at		
25° C (77° F)		
Maintenance Mixtures	2224 (500)	
Paving Mixtures	3336 (750)	
Flow, units of 0.25 mm	8	16
(0.01 in.)		
Percent Stability Retention		
After 4 days in water at	75	
25° C (77° F)		

H.18 SELECTION OF MIX DESIGN.—Whether the paving mixture will be satisfactory is determined by comparing the test properties with the mix design criteria of Table H-1. There must be a range of asphalt contents in which all the design criteria are met. If this range is very narrow (less than 0.5 percent) or if any of the criteria are not met at any asphalt content, adjust the mix by either changing the proportions of coarse and fine aggregates or by substituting another source of aggregate for one of the fractions. In extreme cases, an entirely different source of aggregate may be required.

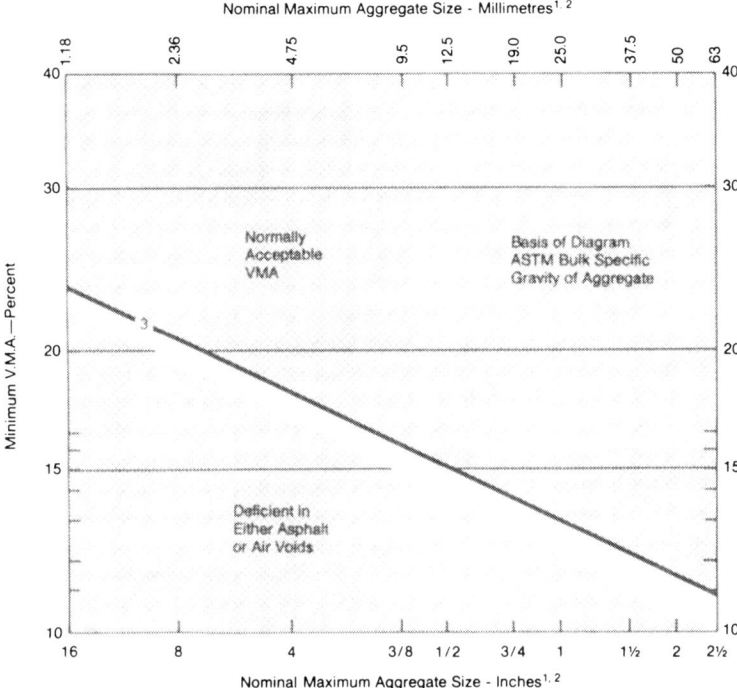

[1]Standard Specification for *Wire Cloth Sieves for Testing Purposes*, ASTM Designation E11 (AASHTO Designation M92).
[2]For processed aggregate, the nominal maximum particle size is the largest sieve size listed in the applicable specification upon which any material is retained.
[3]Mixtures in the 1% tolerance band shall be permitted only when experience indicates that the mixture will perform satisfactorily and when all other criteria are met.
Enter the chart with normal maximum particle size and move vertically to the heavy line. Select a minimum VMA value at or above this line.

Figure H-5. Minimum percent voids in mineral aggregate (VMA).

EXAMPLE

Characteristics of Components—Assume that an aggregate having the characteristics shown in Table H-2 is to be evaluated for a road mix prepared with an MC-250 cutback asphalt whose analysis, viscosity-temperature relationship and composition are defined in Table H-3, Figure H-1, and Figure H-2, respectively. The aggregate gradation plotted in Figure H-6 indicates that it deviates sufficiently from the maximum density curve to meet the minimum requirement of 15 percent VMA for a maximum particle size of 12.7 mm (1/2 in.).

Mixing Temperature—For the preparation of the test specimens, the mixing temperature is first determined from the viscosity-temperature relationship (Figure H-1). The temperature corresponding to a viscosity of 170 ± 20 centistokes is 73.9° C (165° F) and, therefore, is the mixing temperature.

Table H-2. Characteristics of Mineral Aggregate

Gradation: U.S. Sieve Series	
Total percent passing	
12.7 mm (1/2 in.)	100
9.25 mm (3/8 in.)	75.4
4.76 mm (No. 4)	60.7
2.38 mm (No. 8)	55.2
1.19 mm (No. 16)	43.2
0.59 mm (No. 30)	31.6
0.2967 mm (No. 50)	13.2
0.149 mm (No. 100)	5.4
0.074 mm (No. 200)	2.3
Specific Gravity	
Bulk	2.724
Effective[1]	2.757
Apparent	2.796
Water absorption, percent weight	0.97
Asphalt absorption, percent weight	0.45

[1] Determined by ASTM D 2041 using 85-100 penetration asphalt.

Table H-3. Analysis of Medium Curing Asphalts[1]

Grade	MC-250	MC-800	MC-3000
Composition[2]			
MC Cutback Asphalt Base, wt %	78	85	91
MC Cutback Solvent, wt %	22	15	9
Specific Gravity			
at 15.6° C (60° F)	0.962	0.977	0.992
at 25° C (77° F)[3]			
Viscosity at 60° C (140° F), centistokes	359	1211	5150
Distillation			
Percent of Total Distillate to 360° C (680° F)			
to 190° C (374° F)	0	0	0
225° C (437° F)	2.3	0	0
260 ° C (500° F)	40.2	18.2	0
315.5° C (600° F)	80.5	70.9	50.0
Residue from Distillation			
Volume percent by difference	78.3	86.3	93.3
Penetration 100g/5sec/25° C (77° F)	190	186	173
Ductility 5 cm/min/25° C (77° F)	150+	130	123

[1] All grades made from the same basestocks
[2] See Figure H-2 for percentages of solvent corresponding to various viscosities at 60° C (140° F)
[3] See Figure H-3 for specific gravities at 25° C (77° F) corresponding to various solvent contents

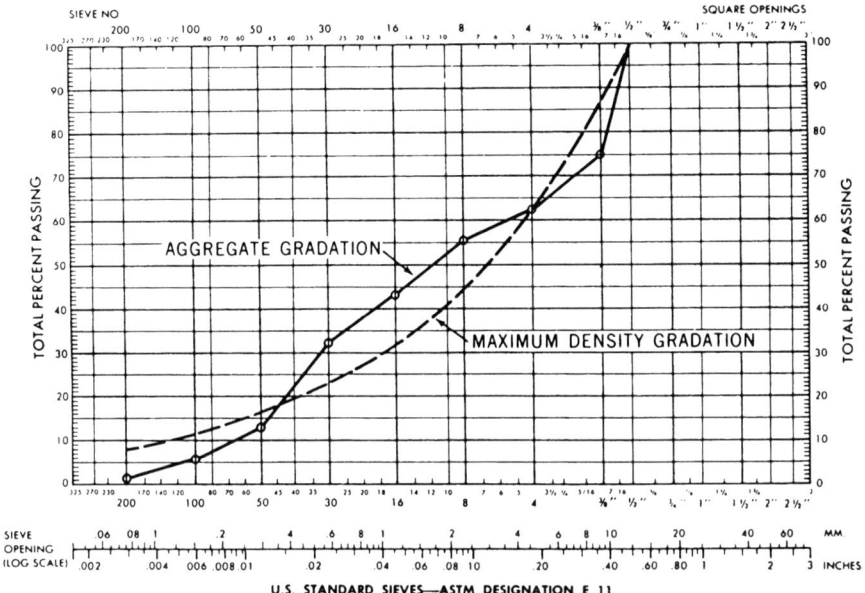

Figure H-6. Example of aggregate gradation showing deviation from maximum density curve.

Compaction Temperature—To determine the compaction temperature, reference is first made to the composition chart for cutback asphalts made from the same base stocks as the MC-250 being used. The viscosity of the MC-250 is 359 centistokes at 60° C (140° F) (Table H-3) and the solvent content read off the chart (Figure H-2) is 22 percent by weight. Its viscosity after losing 50 percent of its solvent will also be read off the chart (Figure H-2) for a solvent content of $22 \times \frac{50}{100} = 11$ percent. This viscosity is 3200 centistokes at 60° C (140° F). The compaction temperature is that at which the viscosity of the cured asphalt is 280 ± 30 centistokes. Since the slopes of the viscosity temperature lines for asphalts made from the same base stocks are practically parallel when the amount of solvent (viscosity) is varied, the compaction temperature for the MC-250 cured to a viscosity of 3200 centistokes at 60° C (140° F) can be obtained by referring to Figure H-1. This temperature is 99° C (210° F). C (165° F) and, therefore, is the mixing temperature.

Curing Temperature—The curing temperature will be the compaction temperature 99° C (210° F) plus 11° C (20° F) to allow for heat loss during the weighing for solvent loss and transfer to the compaction mold. Besides permitting the control of the solvent loss during the heating to the specified compaction temperature, the high curing temperature has the added advantage that it reduces the curing time which, depending on the volatility of the solvent, can be as short as 30 minutes for medium curing asphalts.

Test Data—Test data obtained at five different asphalt contents and representing the average values of three specimens in each series of mixes are shown in Table H-4 and are plotted in Figure H-4.

Table H-4. Characteristics of Paving Mixtures Prepared With MC-250

Asphalt Content (incorporated)						
Kilograms (lbs) per 45.4 kg (100 lb.) of aggregate	2.5(5.5)	2.7(6.0)	2.9(6.5)	3.2(7.0)	3.6(8.0)	
Kilograms (lbs) per 45.4 kg (100 lb.) of mix	2.36(5.21)	2.57(5.66)	2.77(6.10)	2.97(6.54)	3.36(7.41)	
Binder Content (after 50% cure)						
Kilograms (lbs) per 45.4 kg (100 lbs) of aggregate	2.22(4.89)	2.42(5.34)	2.62(5.78)	2.82(6.23)	3.23(7.12)	
Kilograms (lbs) per 45.4 kg (100 lbs) of mix	2.10(4.64)	2.29(5.04)	2.46(5.43)	2.64(5.82)	2.99(6.59)	
Maximum Specific Gravity	2.543	2.526	2.510	2.494	2.46	
Bulk Specific Gravity, test specimens, average	2.373	2.391	2.402	2.413	2.42	
Unit Weight, compacted mix, kg/m^3 (lb/ft^3)	2,382.3(148.1)	2,390(149.2)	2,401.2(149.9)	2,412.4(150.6)	2,422(151.2)	
Air Voids, compacted mix, percent	6.7	5.3	4.3	3.2	1.6	
Voids mineral aggregate, percent	16.9	16.7	16.6	16.6	16.9	
Marshall Stability, (N) lb at 25°C (77°F)	5,338(1,200)	6,036(1,375)	6,228(1,400)	6,450(1,450)	6,005(1,350)	
Marshall Flow at 25°C (77°F), units of 0.25 mm (0.01 in.)	9.0	8.0	8.0	8.5	11.5	
Marshall Stability, lb (N) at 25°C (77°F) after 4-day immersion			5,783(1,300)			
Marshall Flow at 25°C (77°F) after 4-day immersion			8.5			
Marshall Stability Retention, percent			92.8			

Analysis of Data—At 4 percent air voids, all the design criteria of Table H-4 are met:

Asphalt Content, percent by weight of mix	6.2
Unit weight, kg/m^3 (lb/ft^3)	2404.4 (150.1)
Voids mineral aggregate (VMA), percent	16.4
Marshall Stability, N (lb) at 25° C (77° F)	6339 (1,425)
Marshall Flow at 25° C (77° F), (0.01 in.)	8.0
Retained Marshall Stability at 25° C (77° F)* percent	93

*After immersion in water at 25° C (77° F) for 4 days

It will be noted that the asphalt content of 6.2 percent can be varied by ± 0.4 percent without any of the test properties failing to meet the design criteria. Therefore, the asphalt content of 6.2 percent by weight of total mix will be chosen as the optimum asphalt content.

Selection of Mix Design—The Marshall stability at 25° C (77° F) is 6,339 N (1,425 lb) which is well above the minimum of 3,336 N (750 lb) and the flow is at the lower limit of 8. This low flow value, which reflects a high internal friction provided by the aggregate, indicates that the mix should resist deformation during its early life, a critical period for any pavement built with cutback asphalt. The stability could probably be increased appreciably by increasing the percentage of coarse aggregate in the mix. This would be possible because the VMA are much higher than the minimum specified. Increasing the percentage of coarse aggregate (retained on No. 4 or No. 8 sieve) would, therefore, increase the stability without affecting flow, decrease the VMA, and lower the optimum asphalt content. In such a case, the mix would have to be redesigned to determine the actual test properties.

Appendix I. Miscellaneous Tables
Usage Temperatures
Table I-1. Typical Temperatures for Uses of Cutback and Emulsified Asphalt—Degrees Celsius (° C)

Type and Grade of Asphalt	Pugmill Mixture Temperatures [1]		Spraying Temperatures [4]	
	Dense-Graded Mixes	Open-Graded Mixes	Road Mixes	Surface Treatments
Cutback Asphalts (MC, SC)[2]				
30 (MC only)	—	—	—	30+
70	—	—	20+	50+
250	55-80	—	40+	75+
800	75-100	—	55+	95+
3000	80-115	—	—	110+
Emulsified Asphalts				
RS-1	—		—	20-60
RS-2	—		—	50-85
MS-1	10-70[3]		20-70	20-70
MS-2	10-70[3]		20-70	—
MS-2h	10-70[3]		20-70	—
HFMS-1	10-70[3]		20-70	20-70
HFMS-1h	10-70[3]		20-70	—
HFMS-2h	10-70[3]		20-70	—
HFMS-2s	10-70[3]		20-70	—
SS-1	10-70[3]		20-70	—
SS-1h	10-70[3]		20-70	—
CRS-1	—		—	50-85
CRS-2	—		—	50-85
CMS-2	10-70[3]		20-70	—
CMS-2h	10-70[3]		20-70	—
CSS-1	10-70[3]		20-70	—
CSS-1h	10-70[3]		20-70	—

NOTES: Temperatures for cutback asphalts are guides only.

[1] Temperature of mixture immediately after discharge from the pugmill rather than temperature of cutback asphalt.
[2] Application temperature may, in some cases, be above the flash point of the material. Caution must therefore be exercised to prevent fire or an explosion.
[3] Temperature of the emulsified asphalt during mixing.
[4] The maximum temperature (cutback asphalt) shall be below that at which fogging occurs.

Table I-2 Typical Temperatures for Uses of Cutback and Emulsified Asphalt—Degrees Fahrenheit (° F)

Type and Grade of Asphalt	Pugmill Mixture Temperatures [1]		Spraying Temperatures [4]	
	Dense-Graded Mixes	Open-Graded Mixes	Road Mixes	Surface Treatments
Cutback Asphalts (MC, SC)[2]				
30 (MC only)	—	—	—	85+
70	—	—	65+	120+
250	135-175	—	105+	165+
800	165-210	—	135+	200+
3000	180-240	—	—	230+
Emulsified Asphalts				
RS-1	—		—	70-140
RS-2	—		—	125-185
MS-1	50-160[3]		70-160	70-160
MS-2	50-160[3]		70-160	—
MS-2h	50-160[3]		70-160	—
HFMS-1	50-160[3]		70-160	70-160
HFMS-1h	50-160[3]		70-160	—
HFMS-2h	50-160[3]		70-160	—
HFMS-2s	50-160[3]		70-160	—
SS-1	50-160[3]		70-160	—
SS-1h	50-160[3]		70-160	—
CRS-1	—		—	125-185
CRS-2	—		—	125-185
CMS-2	50-160[3]		70-160	—
CMS-2h	50-160[3]		70-160	—
CSS-1	50-160[3]		70-160	—
CSS-1h	50-160[3]		70-160	—

NOTES: Temperatures for cutback asphalts are guides only.

[1] Temperature of mixture immediately after discharge from the pugmill rather than temperature of cutback asphalt.

[2] Application temperature may, in some cases, be above the flash point of the material. Caution must therefore be exercised to prevent fire or an explosion.

[3] Temperature of the emulsified asphalt during mixing.

[4] The maximum temperature (cutback asphalt) shall be below that at which fogging occurs.

CAUTION

The purpose of Tables I-1 and I-2 is to indicate temperature ranges necessary to provide proper asphalt viscosity for spraying and mixing applications for the grades of asphalt shown. It *must* be recognized, however, that temperature ranges indicated by the charts generally are above the minimum flash point for the MC and SC. In fact, some of these cutback asphalts will "flash" at temperatures below these indicated ranges. Accordingly, suitable safety precautions are mandatory at all times when handling these liquid asphalts. These safety precautions include, but are not limited to:

(1) Open flames or sparks must not be permitted close to these materials. Controlled heat should be applied in heating kettles, mixers, distributors, or other equipment designed and approved for the purpose.
(2) Open flames must not be used to inspect or examine drums, tank cars, or other containers in which these materials have been stored.
(3) All vehicles transporting these materials must be properly vented.
(4) Only experienced personnel must be permitted to supervise the handling of these materials.
(5) All applicable intrastate and interstate commerce requirements must be met.

Volume Calculations

I.01 TEMPERATURE-VOLUME RELATIONSHIPS AND CALCULATIONS.—

All liquids and most solids undergo changes in volume with changes in temperature. They expand when heated and contract when cooled. The change in unit volume per degree change in temperature is termed *coefficient of expansion,* a factor that varies with variations in the density (specific gravity) of the asphalt product.

Temperature-volume correction tables for asphalt materials are published in ASTM D4311 and are reproduced here in Tables I-3 (°C) and I-4 (°F). Two columns of correction factors are given in each table. The selection of the appropriate column (A or B) is dependent upon the specific gravity or density of the asphalt, as indicated in the table footnotes. The column A factors apply to the majority of asphalts. Temperature-volume relationships for emulsified asphalts are given in Table I-5.

The multipliers given in the tables are used to convert a known volume at a given temperature to volume at 15° C (60° F), which is customarily used as the standard basis for volume determination of asphalt:

$$V = V_T M_T \qquad (1)$$

where V = volume at 15° C or 60° F,
V_T = volume at given temperature
M_T = multiplier from appropriate table

Example: (U.S. Customary Units)
The specific gravity of an asphalt product is found to be 0.985 at 60° F. The volume of this material is measured to be 9,000 gal at 180° F. The column A factors of Table I-4 apply since the specific gravity exceeds 0.967.
For a temperature of 180° F the correction factor, M_T, is 0.9587. Thus, the volume of the material at 60° F is 9,000 × (0.9587) = 8,628 gal.

I.02 SPECIFIC GRAVITY CALCULATIONS.—

The basic formula for specific gravity of a substance is the mass in air of a unit volume of the substance at a stated temperature divided by the mass in air of an equal volume of gas-free distilled water at a stated temperature. Or:

$$G_x = \frac{W_x}{W_w} \quad (2a) \qquad \text{or} \qquad G_x = \frac{W_x}{V_x \, \gamma_w} \quad (2b)$$

where G_x = specific gravity of the substance
W_x = mass of a unit volume of the substance
W_w = mass of a unit volume of gas-free distilled water
V_x = volume of the substance
γ_w = density of water

The volume of asphalt, which changes with temperature, can be determined using Eq. 1. However, the volume of water (or density) also changes with temperature, as discussed in the following paragraphs.

Table I-3. Temperature-Volume Corrections for Asphalt Materials (Degrees Celsius)

Observed Temperature, °C	Volume Correction[C,D] Factor to 15°C		Observed Temperature, °C	Volume Correction[C,D] Factor to 15°C		Observed Temperature, °C	Volume Correction[C,D] Factor to 15°C		Observed Temperature, °C	Volume Correction[C,D] Factor to 15°C	
	A	B		A	B		A	B		A	B
−25,0	1,0254	1,0290	12,5	1,0016	1,0018	50,0	0,9782	0,9752	87,5	0,9552	0,9492
−24,5	1,0251	1,0286	13,0	1,0012	1,0014	50,5	0,9779	0,9749	88,0	0,9548	0,9489
−24,0	1,0248	1,0283	13,5	1,0009	1,0014	51,0	0,9776	0,9745	88,5	0,9545	0,9485
−23,5	1,0244	1,0279	14,0	1,0006	1,0007	51,5	0,9773	0,9742	89,0	0,9542	0,9482
−23,0	1,0241	1,0276	14,5	1,0003	1,0004	52,0	0,9770	0,9738	89,5	0,9539	0,9478
−22,5	1,0238	1,0272	15,0	1,0000	1,0000	52,5	0,9767	0,9735	90,0	0,9536	0,9475
−22,6	1,0235	1,0268	15,5	0,9997	0,9998	53,0	0,9763	0,9731	90,5	0,9533	0,9472
−21,5	1,0232	1,0265	16,0	0,9994	0,9993	53,5	0,9760	0,9728	91,0	0,9530	0,9468
−21,0	1,0228	1,0261	16,5	0,9991	0,9989	54,0	0,9757	0,9724	91,5	0,9527	0,9465
−20,5	1,0225	1,0258	17,0	0,9988	0,9986	54,5	0,9754	0,9721	92,0	0,9524	0,9461
−20,0	1,0222	1,0254	17,5	0,9985	0,9982	55,0	0,9751	0,9717	92,5	0,9521	0,9458
−19,5	1,0219	1,0250	18,0	0,9981	0,9978	55,5	0,9748	0,9714	93,0	0,9518	0,9455
−19,0	1,0216	1,0247	18,5	0,9978	0,9975	56,0	0,9745	0,9710	93,5	0,9515	0,9451
−18,5	1,0212	1,0243	19,0	0,9975	0,9971	56,5	0,9742	0,9707	94,0	0,9512	0,9448
−18,0	1,0209	1,0239	19,5	0,9972	0,9968	57,0	0,9739	0,9703	94,5	0,9509	0,9444
−17,5	1,0206	1,0236	20,0	0,9969	0,9964	57,5	0,9736	0,9700	95,0	0,9506	0,9441
−17,0	1,0203	1,0232	20,5	0,9966	0,9961	58,0	0,9732	0,9696	95,5	0,9503	0,9438
−16,5	1,0200	1,0228	21,0	0,9963	0,9957	58,5	0,9729	0,9693	96,0	0,9500	0,9434
−16,0	1,0196	1,0224	21,5	0,9959	0,9954	59,0	0,9726	0,9689	96,5	0,9497	0,9431
−15,5	1,0193	1,0221	22,0	0,9956	0,9950	59,5	0,9723	0,9686	97,0	0,9494	0,9427
−15,0	1,0190	1,0217	22,5	0,9953	0,9947	60,0	0,9720	0,9682	97,5	0,9491	0,9424
−14,5	1,0187	1,0213	23,0	0,9950	0,9943	60,5	0,9717	0,9679	98,0	0,9488	0,9421
−14,0	1,0184	1,0210	23,5	0,9947	0,9940	61,0	0,9714	0,9675	98,5	0,9485	0,9417
−13,5	1,0180	1,0206	24,0	0,9943	0,9936	61,5	0,9711	0,9672	99,0	0,9482	0,9414
−13,0	1,0177	1,0203	24,5	0,9940	0,9933	62,0	0,9708	0,9668	99,5	0,9479	0,9410
−12,5	1,0174	1,0199	25,0	0,9937	0,9929	62,5	0,9705	0,9665	100,0	0,9476	0,9407
−12,0	1,0171	1,0195	25,5	0,9937	0,9925	63,0	0,9701	0,9661	100,5	0,9473	0,9404
−11,5	1,0168	1,0192	26,0	0,9934	0,9922	63,5	0,9698	0,9658	101,0	0,9470	0,9400
−11,0	1,0164	1,0188	26,5	0,9928	0,9918	64,0	0,9695	0,9654	101,5	0,9467	0,9397
−10,5	1,0161	1,0185	27,0	0,9925	0,9915	64,5	0,9692	0,9651	102,0	0,9464	0,9393
−10,0	1,0158	1,0181	27,5	0,9922	0,9911	65,0	0,9689	0,9647	102,5	0,9461	0,9390
−9,5	1,0155	1,0177	28,0	0,9918	0,9907	65,5	0,9686	0,9644	103,0	0,9458	0,9387
−9,0	1,0152	1,0174	28,5	0,9915	0,9904	66,0	0,9683	0,9640	103,5	0,9455	0,9383
−8,5	1,0148	1,0170	29,0	0,9912	0,9900	66,5	0,9680	0,9637	104,0	0,9452	0,9380
−8,0	1,0145	1,0166	29,5	0,9909	0,9897	67,0	0,9677	0,9633	104,5	0,9449	0,9376
−7,5	1,0142	1,0163	30,0	0,9906	0,9893	67,5	0,9674	0,9630	105,0	0,9446	0,9373
−7,0	1,0139	1,0159	30,5	0,9903	0,9889	68,0	0,9670	0,9626	105,5	0,9443	0,9370
−6,5	1,0136	1,0155	31,0	0,9900	0,9886	68,5	0,9667	0,9623	106,0	0,9440	0,9366
−6,0	1,0132	1,0151	31,5	0,9897	0,9882	69,0	0,9664	0,9619	106,5	0,9437	0,9363
−5,5	1,0129	1,0148	32,0	0,9894	0,9879	69,5	0,9661	0,9616	107,0	0,9434	0,9359
−5,0	1,0126	1,0144	32,5	0,9891	0,9875	70,0	0,9568	0,9612	107,5	0,9431	0,9356
−4,5	1,0123	1,0140	33,0	0,9887	0,9871	70,5	0,9655	0,9609	108,0	0,9428	0,9353
−4,0	1,0120	1,0137	33,5	0,9984	0,9868	71,0	0,9652	0,9605	108,5	0,9425	0,9349
−3,5	1,0117	1,0133	34,0	0,9881	0,9864	71,5	0,9649	0,9602	109,0	0,9422	0,9346
−3,0	1,0114	1,0130	34,5	0,9878	0,9861	72,0	0,9646	0,9598	109,5	0,9419	0,9342
−2,5	1,0111	1,0126	35,0	0,9875	0,9857	72,5	0,9643	0,9595	110,0	0,9416	0,9339
−2,0	1,0107	1,0122	35,5	0,9872	0,9854	73,0	0,9640	0,9592	110,5	0,9413	0,9336
−1,5	1,0104	1,0119	36,0	0,9869	0,9850	73,5	0,9637	0,9588	111,0	0,9410	0,9932
−1,0	1,0101	1,0115	36,5	0,9866	0,9847	74,0	0,9634	0,9585	111,5	0,9407	0,9329
−0,5	1,0098	1,0112	37,0	0,9863	0,9843	74,5	0,9631	0,9581	112,0	0,9404	0,9325
0	1,0095	1,0108	37,5	0,9860	0,9840	75,0	0,9628	0,9578	112,5	0,9401	0,9322
0,5	1,0092	1,0104	38,0	0,9856	0,9836	75,5	0,9625	0,9575	113,0	0,9397	0,9319
1,0	1,0089	1,0101	38,5	0,9853	0,9833	76,0	0,9622	0,9571	113,5	0,9394	0,9315
1,5	1,0085	1,0097	39,0	0,9850	0,9829	76,5	0,9619	0,9568	114,0	0,9391	0,9312
2,0	1,0082	1,0094	39,5	0,9847	0,9826	77,0	0,9616	0,9564	114,5	0,9388	0,9308
2,5	1,0079	1,0090	40,0	0,9844	0,9822	77,5	0,9613	0,9561	115,0	0,9385	0,9305
3,0	1,0076	1,0086	40,5	0,9841	0,9819	78,0	0,9609	0,9557	115,5	0,9382	0,9302
3,5	1,0073	1,0083	41,0	0,9838	0,9815	78,5	0,9606	0,9554	116,0	0,9379	0,9298
4,0	1,0069	1,0079	41,5	0,9835	0,9812	79,0	0,9603	0,9550	116,5	0,9376	0,9295
4,5	1,0066	1,0076	42,0	0,9832	0,9808	79,5	0,9600	0,9547	117,0	0,9373	0,9292
5,0	1,0063	1,0072	42,5	0,9829	0,9805	80,0	0,9597	0,9543	117,5	0,9371	0,9289
5,5	1,0060	1,0068	43,0	0,9825	0,9801	80,5	0,9594	0,9540	118,0	0,9368	0,9285
6,0	1,0057	1,0065	43,5	0,9822	0,9798	81,0	0,9591	0,9536	118,5	0,9365	0,9282
6,5	1,0053	1,0061	44,0	0,9819	0,9794	81,5	0,9588	0,9533	119,0	0,9362	0,9279
7,0	1,0050	1,0058	44,5	0,9816	0,9791	82,0	0,9585	0,9529	119,5	0,9359	0,9275
7,5	1,0047	1,0054	45,0	0,9813	0,9787	82,5	0,9582	0,9526	120,0	0,9356	0,9272
8,0	1,0044	1,0050	45,5	0,9810	0,9784	83,0	0,9578	0,9523	120,5	0,9353	0,9269
8,5	1,0041	1,0047	46,0	0,9807	0,9780	83,5	0,9576	0,9519	121,0	0,9350	0,9265
9,0	1,0037	1,0043	46,5	0,9804	0,9777	84,0	0,9573	0,9516	121,5	0,9347	0,9262
9,5	1,0034	1,0040	47,0	0,9801	0,9773	84,5	0,9570	0,9512	122,0	0,9344	0,9258
10,0	1,0031	1,0036	47,5	0,9798	0,9770	85,0	0,9567	0,9509	122,5	0,9341	0,9255
10,5	1,0028	1,0032	48,0	0,9794	0,9766	85,5	0,9564	0,9506	123,0	0,9338	0,9252
11,0	1,0025	1,0029	48,5	0,9791	0,9763	86,0	0,9561	0,9502	123,5	0,9335	0,9248
11,5	1,0022	1,0025	49,0	0,9788	0,9759	86,5	0,9558	0,9499	124,0	0,9332	0,9245
12,0	1,0019	1,0022	49,5	0,9785	0,9756	87,0	0,9555	0,9495	124,5	0,9329	0,9241

[C] Use column A factors for asphalts with density at 15°C of 966 kg/m³ or higher.
[D] Use column B factors for asphalts with density at 15°C of 850 to 965 kg/m³.

Table I-3. (Contd.) Temperature-Volume Corrections for Asphalt Materials (Degrees Celsius)

Observed Temperature, °C	Volume Correction[C,D] Factor to 15°C A	B	Observed Temperature, °C	Volume Correction[C,D] Factor to 15°C A	B	Observed Temperature, °C	Volume Correction[C,D] Factor to 15°C A	B	Observed Temperature, °C	Volume Correction[C,D] Factor to 15°C A	B
125,0	0,9326	0,9238	162,5	0,9104	0,8991	200,0	0,8886	0,8749	237,5	0,8673	0,8514
125,5	0,9323	0,9235	163,0	0,9101	0,8987	200,5	0,8883	0,8746	238,0	0,8670	0,8510
126,0	0,9320	0,9231	163,5	0,9098	0,8984	201,0	0,8880	0,8743	238,5	0,8667	0,8507
126,5	0,9317	0,9228	164,0	0,9095	0,8981	201,5	0,8877	0,8739	239,0	0,8664	0,8504
127,0	0,9314	0,9225	164,5	0,9092	0,8977	202,0	0,8874	0,8736	239,5	0,8661	0,8501
127,5	0,9311	0,9222	165,0	0,9089	0,8974	202,5	0,8872	0,8733	240,0	0,8658	0,8498
128,0	0,9308	0,9218	165,5	0,9086	0,8971	203,0	0,8869	0,8730	240,5	0,8655	0,8495
128,5	0,9305	0,9215	166,0	0,9083	0,8968	203,5	0,8866	0,8727	241,0	0,8652	0,8492
129,0	0,9302	0,9212	166,5	0,9080	0,8964	204,0	0,8863	0,8723	241,5	0,8650	0,8489
129,5	0,9299	0,9208	167,0	0,9077	0,8961	204,5	0,8860	0,8720	242,0	0,8647	0,8486
130,0	0,9296	0,9205	167,5	0,9075	0,8958	205,0	0,8857	0,8717	242,5	0,8644	0,8483
130,5	0,9293	0,9202	168,0	0,9072	0,8955	205,5	0,8854	0,8714	243,0	0,8641	0,8480
131,0	0,9290	0,9198	168,5	0,9069	0,8952	206,0	0,8851	0,8711	243,5	0,8638	0,8477
131,5	0,9287	0,9195	169,0	0,9066	0,8948	206,5	0,8849	0,8708	244,0	0,8636	0,8474
132,0	0,9284	0,9191	169,5	0,9063	0,8945	207,0	0,8846	0,8705	244,5	0,8633	0,8471
132,5	0,9281	0,9188	170,0	0,9060	0,8942	207,5	0,8843	0,8702	245,0	0,8630	0,8468
133,0	0,9278	0,9185	170,5	0,9057	0,8939	208,0	0,8840	0,8698	245,5	0,8627	0,8465
133,5	0,9275	0,9181	171,0	0,9054	0,8935	208,5	0,8837	0,8695	246,0	0,8624	0,8462
134,0	0,9272	0,9178	171,5	0,9051	0,8932	209,0	0,8835	0,8692	246,5	0,8622	0,8459
134,5	0,9269	0,9174	172,0	0,9048	0,8929	209,5	0,8832	0,8689	247,0	0,8619	0,8456
135,0	0,9266	0,9171	172,5	0,9046	0,8926	210,0	0,8829	0,8686	247,5	0,8616	0,8453
135,5	0,9263	0,9168	173,0	0,9043	0,8922	210,5	0,8826	0,8683	248,0	0,8613	0,8449
136,0	0,9260	0,9164	173,5	0,9040	0,8919	211,0	0,8823	0,8680	248,5	0,8610	0,8446
136,5	0,9257	0,9161	174,0	0,9037	0,8916	211,5	0,8820	0,8676	249,0	0,8608	0,8443
137,0	0,9254	0,9158	174,5	0,9034	0,8912	212,0	0,8817	0,8673	249,5	0,8605	0,8440
137,5	0,9251	0,9155	175,0	0,9031	0,8909	212,5	0,8815	0,8670	250,0	0,8602	0,8437
138,0	0,9248	0,9151	175,5	0,9028	0,8906	213,0	0,8812	0,8667	250,5	0,8599	0,8434
138,5	0,9246	0,9148	176,0	0,9025	0,8903	213,5	0,8809	0,8664	251,0	0,8596	0,8431
139,0	0,9242	0,9145	176,5	0,9022	0,8899	214,0	0,8806	0,8660	251,5	0,8594	0,8428
139,5	0,9239	0,9141	177,0	0,9019	0,8896	214,5	0,8803	0,8657	252,0	0,8591	0,8425
140,0	0,9236	0,9138	177,5	0,9017	0,8893	215,0	0,8800	0,8654	252,5	0,8588	0,8422
140,5	0,9233	0,9135	178,0	0,9014	0,8890	215,5	0,8797	0,8651	253,0	0,8585	0,8418
141,0	0,9230	0,9131	178,5	0,9011	0,8887	216,0	0,8794	0,8648	253,5	0,8582	0,8415
141,5	0,9227	0,9128	179,0	0,9008	0,8883	216,5	0,8792	0,8645	254,0	0,8580	0,8412
142,0	0,9224	0,9125	179,5	0,9005	0,8880	217,0	0,8789	0,8642	254,5	0,8577	0,8409
142,5	0,9222	0,9122	180,0	0,9002	0,8877	217,5	0,8786	0,8639	255,0	0,8574	0,8406
143,0	0,9219	0,9118	180,5	0,8999	0,8874	218,0	0,8783	0,8635	255,5	0,8571	0,8403
143,5	0,9216	0,9115	181,0	0,8996	0,8871	218,5	0,8780	0,8632	256,0	0,8568	0,8400
144,0	0,9213	0,9112	181,5	0,8993	0,8867	219,0	0,8778	0,8629	256,5	0,8566	0,8397
144,5	0,9210	0,9108	182,0	0,8990	0,8864	219,5	0,8775	0,8626	257,0	0,8563	0,8394
145,0	0,9207	0,9105	182,5	0,8988	0,8861	220,0	0,8772	0,8623	257,5	0,8560	0,8391
145,5	0,9204	0,9102	183,0	0,8985	0,8858	220,5	0,8769	0,8620	258,0	0,8557	0,8388
146,0	0,9201	0,9098	183,5	0,8982	0,8855	221,0	0,8766	0,8617	258,5	0,8554	0,8385
146,5	0,9198	0,9095	184,0	0,8979	0,8851	221,5	0,8763	0,8614	259,0	0,8552	0,8382
147,0	0,9195	0,9092	184,5	0,8976	0,8848	222,0	0,8760	0,8611	259,5	0,8549	0,8379
147,5	0,9192	0,9089	185,0	0,8973	0,8845	222,5	0,8758	0,8608	260,0	0,8546	0,8376
148,0	0,9189	0,9085	185,5	0,8970	0,8842	223,0	0,8755	0,8604	260,5	0,8543	0,8373
148,5	0,9186	0,9082	186,0	0,8967	0,8839	223,5	0,8752	0,8601	261,0	0,8540	0,8370
149,0	0,9183	0,9079	186,5	0,8964	0,8835	224,0	0,8749	0,8598	261,5	0,8538	0,8367
149,5	0,9180	0,9075	187,0	0,8961	0,8832	224,5	0,8746	0,8595	262,0	0,8535	0,8364
150,0	0,9177	0,9072	187,5	0,8959	0,8829	225,0	0,8743	0,8592	262,5	0,8532	0,8361
150,5	0,9174	0,9069	188,0	0,8956	0,8826	225,5	0,8740	0,8589	263,0	0,8529	0,8357
151,0	0,9171	0,9065	188,5	0,8953	0,8823	226,0	0,8737	0,8586	263,5	0,8526	0,8354
151,5	0,9168	0,9062	189,0	0,8950	0,8819	226,5	0,8735	0,8582	264,0	0,8524	0,8351
152,0	0,9165	0,9059	189,5	0,8947	0,8816	227,0	0,8732	0,8579	264,5	0,8521	0,8348
152,5	0,9163	0,9056	190,0	0,8944	0,8813	227,5	0,8729	0,8576	265,0	0,8518	0,8345
153,0	0,9160	0,9052	190,5	0,8941	0,8810	228,0	0,8726	0,8573	265,5	0,8515	0,8342
153,5	0,9157	0,9049	191,0	0,8938	0,8807	228,5	0,8723	0,8570	266,0	0,8512	0,8339
154,0	0,9154	0,9046	191,5	0,8935	0,8803	229,0	0,8721	0,8566	266,5	0,8510	0,8336
154,5	0,9151	0,9042	192,0	0,8932	0,8800	229,5	0,8718	0,8563	267,0	0,8507	0,8333
155,0	0,9148	0,9039	192,5	0,8930	0,8797	230,0	0,8715	0,8560	267,5	0,8504	0,8330
155,5	0,9145	0,9036	193,0	0,8927	0,8794	230,5	0,8712	0,8557	268,0	0,8501	0,8326
156,0	0,9142	0,9033	193,5	0,8924	0,8791	231,0	0,8709	0,8554	268,5	0,8498	0,8323
156,5	0,9139	0,9029	194,0	0,8921	0,8787	231,5	0,8707	0,8551	269,0	0,8496	0,8320
157,0	0,9136	0,9026	194,5	0,8918	0,8784	232,0	0,8704	0,8548	269,5	0,8493	0,8317
157,5	0,9133	0,9023	195,0	0,8915	0,8781	232,5	0,8701	0,8545	270,0	0,8490	0,8314
158,0	0,9130	0,9020	195,5	0,8912	0,8778	233,0	0,8698	0,8541	270,5	0,8487	0,8311
158,5	0,9127	0,9017	196,0	0,8909	0,8775	233,5	0,8695	0,8538	271,0	0,8484	0,8308
159,0	0,9124	0,9013	196,5	0,8906	0,8771	234,0	0,8693	0,8535	271,5	0,8482	0,8305
159,5	0,9121	0,9010	197,0	0,8903	0,8768	234,5	0,8690	0,8532	272,0	0,8479	0,8302
160,0	0,9118	0,9007	197,5	0,8901	0,8765	235,0	0,8687	0,8529	272,5	0,8476	0,8299
160,5	0,9115	0,9004	198,0	0,8898	0,8762	235,5	0,8684	0,8526	273,0	0,8473	0,8296
161,0	0,9112	0,9000	198,5	0,8895	0,8759	236,0	0,8681	0,8523	273,5	0,8470	0,8293
161,5	0,9109	0,8997	199,0	0,8892	0,8755	236,5	0,8678	0,8520	274,0	0,8468	0,8290
162,0	0,9106	0,8994	199,5	0,8889	0,8752	237,0	0,8675	0,8517	274,5	0,8465	0,8287

[C] Use column A factors for asphalts with density at 15°C of 966 kg/m³ or higher.
[D] Use column B factors for asphalts with density at 15°C of 850 to 965 kg/m³.

Table I-4. Temperature-Volume Corrections for Asphalt Materials (Degrees Fahrenheit)

Observed Temperature, °F	Volume Correction Factor to 60°F [A,B]		Observed Temperature, °F	Volume Correction Factor to 60°F [A,B]		Observed Temperature, °F	Volume Correction Factor to 60°F [A,B]		Observed Temperature, °F	Volume Correction Factor to 60°F [A,B]	
	A	B		A	B		A	B		A	B
0	1.0211	1.0241	75	0.9948	0.9940	150	0.9689	0.9647	225	0.9436	0.9361
1	1.0208	1.0237	76	0.9944	0.9936	151	0.9686	0.9643	226	0.9432	0.9358
2	1.0204	1.0233	77	0.9941	0.9932	152	0.9682	0.9639	227	0.9429	0.9354
3	1.0201	1.0229	78	0.9937	0.9929	153	0.9679	0.9635	228	0.9426	0.9350
4	1.0197	1.0225	79	0.9934	0.9925	154	0.9675	0.9632	229	0.9422	0.9346
5	1.0194	1.0221	80	0.9930	0.9921	155	0.9672	0.9628	230	0.9419	0.9343
6	1.0190	1.0217	81	0.9927	0.9917	156	0.9669	0.9624	231	0.9416	0.9339
7	1.0186	1.0213	82	0.9923	0.9913	157	0.9665	0.9620	232	0.9412	0.9335
8	1.0183	1.0209	83	0.9920	0.9909	158	0.9662	0.9616	233	0.9409	0.9331
9	1.0179	1.0205	84	0.9916	0.9905	159	0.9658	0.9612	234	0.9405	0.9328
10	1.0176	1.0201	85	0.9913	0.9901	160	0.9655	0.9609	235	0.9402	0.9324
11	1.0172	1.0197	86	0.9909	0.9897	161	0.9652	0.9605	236	0.9399	0.9320
12	1.0169	1.0193	87	0.9906	0.9893	162	0.9648	0.9601	237	0.9395	0.9316
13	1.0165	1.0189	88	0.9902	0.9889	163	0.9645	0.9597	238	0.9392	0.9313
14	1.0162	1.0185	89	0.9899	0.9885	164	0.9641	0.9593	239	0.9389	0.9309
15	1.0158	1.0181	90	0.9896	0.9881	165	0.9638	0.9589	240	0.9385	0.9305
16	1.0155	1.0177	91	0.9892	0.9877	166	0.9635	0.9585	241	0.9382	0.9301
17	1.0151	1.0173	92	0.9889	0.9873	167	0.9631	0.9582	242	0.9379	0.9298
18	1.0148	1.0168	93	0.9885	0.9869	168	0.9628	0.9578	243	0.9375	0.9294
19	1.0144	1.0164	94	0.9882	0.9865	169	0.9624	0.9574	244	0.9372	0.9290
20	1.0141	1.0160	95	0.9878	0.9861	170	0.9621	0.9570	245	0.9369	0.9286
21	1.0137	1.0156	96	0.9875	0.9857	171	0.9618	0.9566	246	0.9365	0.9283
22	1.0133	1.0152	97	0.9871	0.9854	172	0.9614	0.9562	247	0.9362	0.9279
23	1.0130	1.0148	98	0.9868	0.9850	173	0.9611	0.9559	248	0.9359	0.9275
24	1.0126	1.0144	99	0.9864	0.9846	174	0.9607	0.9555	249	0.9356	0.9272
25	1.0123	1.0140	100	0.9861	0.9842	175	0.9604	0.9551	250	0.9352	0.9268
26	1.0119	1.0136	101	0.9857	0.9838	176	0.9601	0.9547	251	0.9349	0.9264
27	1.0116	1.0132	102	0.9854	0.9834	177	0.9597	0.9543	252	0.9346	0.9260
28	1.0112	1.0128	103	0.9851	0.9830	178	0.9594	0.9539	253	0.9342	0.9257
29	1.0109	1.0124	104	0.9847	0.9826	179	0.9590	0.9536	254	0.9339	0.9253
30	1.0105	1.0120	105	0.9844	0.9822	180	0.9587	0.9532	255	0.9336	0.9249
31	1.0102	1.0116	106	0.9840	0.9818	181	0.9584	0.9528	256	0.9332	0.9245
32	1.0098	1.0112	107	0.9887	0.9814	182	0.9580	0.9524	257	0.9329	0.9242
33	1.0095	1.0108	108	0.9833	0.9810	183	0.9577	0.9520	258	0.9326	0.9238
34	1.0091	1.0104	109	0.9830	0.9806	184	0.9574	0.9517	259	0.9322	0.9234
35	1.0088	1.0100	110	0.9826	0.9803	185	0.9570	0.9513	260	0.9319	0.9231
36	1.0084	1.0096	111	0.9823	0.9799	186	0.9567	0.9509	261	0.9316	0.9227
37	1.0081	1.0092	112	0.9819	0.9795	187	0.9563	0.9505	262	0.9312	0.9223
38	1.0077	1.0088	113	0.9816	0.9791	188	0.9560	0.9501	263	0.9309	0.9219
39	1.0074	1.0084	114	0.9813	0.9787	189	0.9557	0.9498	264	0.9306	0.9216
40	1.0070	1.0080	115	0.9809	0.9783	190	0.9553	0.9494	265	0.9302	0.9212
41	1.0067	1.0076	116	0.9806	0.9779	191	0.9550	0.9490	266	0.9299	0.9208
42	1.0063	1.0072	117	0.9802	0.9775	192	0.9547	0.9486	267	0.9296	0.9205
43	1.0060	1.0068	118	0.9799	0.9771	193	0.9543	0.9482	268	0.9293	0.9201
44	1.0056	1.0064	119	0.9795	0.9767	194	0.9540	0.9478	269	0.9289	0.9197
45	1.0053	1.0060	120	0.9792	0.9763	195	0.9536	0.9475	270	0.9286	0.9194
46	1.0040	1.0056	121	0.9788	0.9760	196	0.9533	0.9471	271	0.9283	0.9190
47	1.0046	1.0052	122	0.9785	0.9756	197	0.9530	0.9467	272	0.9279	0.9186
48	1.0042	1.0048	123	0.9782	0.9752	198	0.9526	0.9463	273	0.9276	0.9182
49	1.0038	1.0044	124	0.9778	0.9748	199	0.9523	0.9460	274	0.9273	0.9179
50	1.0035	1.0040	125	0.9775	0.9744	200	0.9520	0.9456	275	0.9269	0.9175
51	1.0031	1.0036	126	0.9771	0.9740	201	0.9516	0.9452	276	0.9266	0.9171
52	1.0028	1.0032	127	0.9768	0.9736	202	0.9513	0.9448	277	0.9263	0.9168
53	1.0024	1.0028	128	0.9764	0.9732	203	0.9509	0.9444	278	0.9259	0.9164
54	1.0021	1.0024	129	0.9761	0.9728	204	0.9506	0.9441	279	0.9256	0.9160
55	1.0017	1.0020	130	0.9758	0.9725	205	0.9503	0.9437	280	0.9253	0.9157
56	1.0014	1.0016	131	0.9754	0.9721	206	0.9499	0.9433	281	0.9250	0.9153
57	1.0010	1.0012	132	0.9751	0.9717	207	0.9496	0.9429	282	0.9246	0.9149
58	1.0007	1.0008	133	0.9747	0.9713	208	0.9493	0.9425	283	0.9243	0.9146
59	1.0003	1.0004	134	0.9744	0.9709	209	0.9489	0.9422	284	0.9240	0.9142
60	1.0000	1.0000	135	0.9740	0.9705	210	0.9486	0.9418	285	0.9236	0.9138
61	0.9997	0.9996	136	0.9737	0.9701	211	0.9483	0.9414	286	0.9233	0.9135
62	0.9993	0.9992	137	0.9734	0.9697	212	0.9479	0.9410	287	0.9230	0.9131
63	0.9990	0.9988	138	0.9730	0.9693	213	0.9476	0.9407	288	0.9227	0.9127
64	0.9986	0.9984	139	0.9727	0.9690	214	0.9472	0.9403	289	0.9223	0.9124
65	0.9983	0.9980	140	0.9723	0.9686	215	0.9469	0.9399	290	0.9220	0.9120
66	0.9979	0.9976	141	0.9720	0.9682	216	0.9466	0.9395	291	0.9217	0.9116
67	0.9976	0.9972	142	0.9716	0.9678	217	0.9462	0.9391	292	0.9213	0.9113
68	0.9972	0.9968	143	0.9713	0.9674	218	0.9459	0.9388	293	0.9210	0.9109
69	0.9969	0.9964	144	0.9710	0.9670	219	0.9456	0.9384	294	0.9207	0.9105
70	0.9965	0.9960	145	0.9706	0.9666	220	0.9452	0.9380	295	0.9204	0.9102
71	0.9962	0.9956	146	0.9703	0.9662	221	0.9449	0.9376	296	0.9200	0.9098
72	0.9958	0.9952	147	0.9699	0.9659	222	0.9446	0.9373	297	0.9197	0.9094
73	0.9955	0.9948	148	0.9696	0.9655	223	0.9442	0.9369	298	0.9194	0.9091
74	0.9951	0.9944	149	0.9693	0.9651	224	0.9439	0.9365	299	0.9190	0.9087

[A] Use column A factors for asphalts with API gravity at 60°F of 14.9° or less or with specific gravity 60/60°F of 0.967 or higher.
[B] Use column B factors for asphalts with API gravity at 60°F from 15.0° to 34.9° or with specific gravity 60/60°F from 0.850 to 0.966.

Table I-4. (Contd.) Temperature-Volume Corrections for Asphalt Materials (Degrees Fahrenheit)

Observed Tempera-ture, °F	Volume Correction Factor to 60°F [a,b] A	B	Observed Tempera-ture, °F	Volume Correction Factor to 60°F [a,b] A	B	Observed Tempera-ture, °F	Volume Correction Factor to 60°F [a,b] A	B	Observed Tempera-ture, °F	Volume Correction Factor to 60°F [a,b] A	B
300	0.9187	0.9083	350	0.9024	0.8902	400	0.8864	0.8724	450	0.8705	0.8550
301	0.9184	0.9080	351	0.9021	0.8899	401	0.8861	0.8721	451	0.8702	0.8547
302	0.9181	0.9076	352	0.9018	0.8895	402	0.8857	0.8717	452	0.8699	0.8543
303	0.9177	0.9072	353	0.9015	0.8891	403	0.8854	0.8714	453	0.8696	0.8540
304	0.9174	0.9069	354	0.9011	0.8888	404	0.8851	0.8710	454	0.8693	0.8536
305	0.9171	0.9065	355	0.9008	0.8884	405	0.8848	0.8707	455	0.8690	0.8533
306	0.9167	0.9061	356	0.9005	0.8881	406	0.8845	0.8703	456	0.8687	0.8529
307	0.9164	0.9058	357	0.9002	0.8877	407	0.8841	0.8700	457	0.8683	0.8526
308	0.9161	0.9054	358	0.8998	0.8873	408	0.8838	0.8696	458	0.8680	0.8522
309	0.9158	0.9050	359	0.8995	0.8870	409	0.8835	0.8693	459	0.8677	0.8519
310	0.9154	0.9047	360	0.8992	0.8866	410	0.8832	0.8689	460	0.8674	0.8516
311	0.9151	0.9043	361	0.8989	0.8863	411	0.8829	0.8686	461	0.8671	0.8512
312	0.9148	0.9039	362	0.8986	0.8859	412	0.8826	0.8682	462	0.8668	0.8509
313	0.9145	0.9036	363	0.8982	0.8856	413	0.8822	0.8679	463	0.8665	0.8505
314	0.9141	0.9032	364	0.8979	0.8852	414	0.8819	0.8675	464	0.8661	0.8502
315	0.9138	0.9029	365	0.8976	0.8848	415	0.8816	0.8672	465	0.8658	0.8498
316	0.9135	0.9025	366	0.8973	0.8845	416	0.8813	0.8668	466	0.8655	0.8495
317	0.9132	0.9021	367	0.8969	0.8841	417	0.8810	0.8665	467	0.8652	0.8492
318	0.9128	0.9018	368	0.8966	0.8838	418	0.8806	0.8661	468	0.8649	0.8488
319	0.9125	0.9014	369	0.8963	0.8834	419	0.8803	0.8658	469	0.8646	0.8485
320	0.9122	0.9010	370	0.8960	0.8831	420	0.8800	0.8654	470	0.8643	0.8481
321	0.9118	0.9007	371	0.8957	0.8827	421	0.8797	0.8651	471	0.8640	0.8478
322	0.9115	0.9003	372	0.8953	0.8823	422	0.8794	0.8647	472	0.8636	0.8474
323	0.9112	0.9000	373	0.8950	0.8820	423	0.8791	0.8644	473	0.8633	0.8471
324	0.9109	0.8996	374	0.8947	0.8816	424	0.8787	0.8640	474	0.8630	0.8468
325	0.9105	0.8992	375	0.8944	0.8813	425	0.8784	0.8637	475	0.8627	0.8464
326	0.9102	0.8989	376	0.8941	0.8809	426	0.8781	0.8633	476	0.8624	0.8461
327	0.9099	0.8985	377	0.8937	0.8806	427	0.8778	0.8630	477	0.8621	0.8457
328	0.9096	0.8981	378	0.8934	0.8802	428	0.8775	0.8626	478	0.8618	0.8454
329	0.9092	0.8978	379	0.8931	0.8799	429	0.8772	0.8623	479	0.8615	0.8451
330	0.9089	0.8974	380	0.8928	0.8795	430	0.8768	0.8619	480	0.8611	0.8447
331	0.9086	0.8971	381	0.8924	0.8792	431	0.8765	0.8616	481	0.8608	0.8444
332	0.9083	0.8967	382	0.8921	0.8788	432	0.8762	0.8612	482	0.8605	0.8440
333	0.9079	0.8963	383	0.8918	0.8784	433	0.8759	0.8609	483	0.8602	0.8437
334	0.9076	0.8960	384	0.8915	0.8781	434	0.8756	0.8605	484	0.8599	0.8433
335	0.9073	0.8956	385	0.8912	0.8777	435	0.8753	0.8602	485	0.8596	0.8430
336	0.9070	0.8952	386	0.8908	0.8774	436	0.8749	0.8599	486	0.8593	0.8427
337	0.9066	0.8949	387	0.8905	0.8770	437	0.8746	0.8595	487	0.8590	0.8423
338	0.9063	0.8945	388	0.8902	0.8767	438	0.8743	0.8592	488	0.8587	0.8420
339	0.9060	0.8942	389	0.8899	0.9763	439	0.8740	0.8588	489	0.8583	0.8416
340	0.9057	0.8938	390	0.8896	0.8760	440	0.8737	0.8585	490	0.8580	0.8413
341	0.9053	0.8934	391	0.8892	0.8756	441	0.8734	0.8581	491	0.8577	0.8410
342	0.9050	0.8931	392	0.8889	0.8753	442	0.8731	0.8578	492	0.8574	0.8406
343	0.9047	0.8927	393	0.8886	0.8749	443	0.8727	0.8574	493	0.8571	0.8403
344	0.9044	0.8924	394	0.8883	0.8746	444	0.8724	0.8571	494	0.8568	0.8399
345	0.9040	0.8920	395	0.8880	0.8742	445	0.8721	0.8567	495	0.8565	0.8396
346	0.9037	0.8916	396	0.8876	0.8738	446	0.8718	0.8564	496	0.8562	0.8393
347	0.9034	0.8913	397	0.8873	0.8735	447	0.8715	0.8560	497	0.8559	0.8389
348	0.9031	0.8909	398	0.8870	0.8731	448	0.8712	0.8557	498	0.8556	0.8386
349	0.9028	0.8906	399	0.8867	0.8728	449	0.8709	0.8554	499	0.8552	0.8383
									500	0.8549	0.8379

[a] Use column A factors for asphalts with API gravity at 60°F of 14.9° or less or with specific gravity 60/60°F of 0.967 or higher.
[b] Use column B factors for asphalts with API gravity at 60°F from 15.0° to 34.9° or with specific gravity 60/60°F from 0.850 to 0.966.

Table I-5. Temperature-Volume Corrections for Emulsified Asphalts

Legend: t = observed temperature in degrees Celsius (Fahrenheit)
M = multiplier for correcting volumes to 15° C (60° F)

°C^t	°F	M*	°C^t	°F	M*	°C^t	°F	M*
10.0	50	1.00250	35.0	95	0.99125	60.0	140	0.98000
10.6	51	1.00225	35.6	96	0.99100	60.6	141	0.97975
11.1	52	1.00200	36.1	97	0.99075	61.1	142	0.97950
11.7	53	1.00175	36.7	98	0.99050	61.7	143	0.97925
12.2	54	1.00150	37.2	99	0.99025	62.2	144	0.97900
12.8	55	1.00125	37.8	100	0.99000	62.8	145	0.97875
13.3	56	1.00100	38.3	101	0.98975	63.3	146	0.97850
13.9	57	1.00075	38.9	102	0.98950	63.9	147	0.97825
14.4	58	1.00050	39.4	103	0.98925	64.4	148	0.97800
15.0	59	1.00025	40.0	104	0.98900	65.0	149	0.97775
15.6	60	1.00000	40.6	105	0.98875	65.6	150	0.97750
16.1	61	0.99975	41.1	106	0.98850	66.1	151	0.97725
16.7	62	0.99950	41.7	107	0.98825	66.7	152	0.97700
17.2	63	0.99925	42.2	108	0.98800	67.2	153	0.97675
17.8	64	0.99900	42.8	109	0.98775	67.8	154	0.97650
18.3	65	0.99875	43.3	110	0.98750	68.3	155	0.97625
18.9	66	0.99850	43.9	111	0.98725	68.9	156	0.97600
19.4	67	0.99825	44.4	112	0.98700	69.4	157	0.97575
20.0	68	0.99800	45.0	113	0.98675	70.0	158	0.97550
20.6	69	0.99775	45.6	114	0.98650	70.6	159	0.97525
21.1	70	0.99750	46.1	115	0.98625	71.1	160	0.97500
21.7	71	0.99725	46.7	116	0.98600	71.7	161	0.97475
22.2	72	0.99700	47.2	117	0.98575	72.2	162	0.97450
22.8	73	0.99675	47.8	118	0.98550	72.8	163	0.97425
23.3	74	0.99650	48.3	119	0.98525	73.3	164	0.97400
23.9	75	0.99625	48.9	120	0.98500	73.9	165	0.97375
24.4	76	0.99600	49.4	121	0.98475	74.4	166	0.97350
25.0	77	0.99575	50.0	122	0.98450	75.0	167	0.97325
25.6	78	0.99550	50.6	123	0.98425	75.6	168	0.97300
26.1	79	0.99525	51.1	124	0.98400	76.1	169	0.97275
26.7	80	0.99500	51.7	125	0.98375	76.7	170	0.97250
27.2	81	0.99475	52.2	126	0.98350	77.2	171	0.97225
27.8	82	0.99450	52.8	127	0.98325	77.8	172	0.97200
28.3	83	0.99425	53.3	128	0.98300	78.3	173	0.97175
28.9	84	0.99400	53.9	129	0.98275	78.9	174	0.97150
29.4	85	0.99375	54.4	130	0.98250	79.4	175	0.97125
30.0	86	0.99350	55.0	131	0.98225	80.0	176	0.97100
30.6	87	0.99325	55.6	132	0.98200	80.6	177	0.97075
31.1	88	0.99300	56.1	133	0.98175	81.1	178	0.97050
31.7	89	0.99275	56.7	134	0.98150	81.7	179	0.97025
32.2	90	0.99250	57.2	135	0.98125	82.2	180	0.97000
32.8	91	0.99225	57.8	136	0.98100	82.8	181	0.96975
33.3	92	0.99200	58.3	137	0.98075	83.3	182	0.96950
33.9	93	0.99175	58.9	138	0.98050	83.9	183	0.96925
34.4	94	0.99150	59.4	139	0.98025	84.4	184	0.96900
						85.0	185	0.96875

After the specific gravity of asphalt is found at 25/25° C (77/77° F), its specific gravity at 15.6/15.6° C (60/60° F) can be calculated by expressing Eq. 2b for both temperatures, combining, appropriately substituting Eq. 1 and simplifying to become

$$G_{15.6/15.6} = \frac{G_{25/25} \; \gamma_{w25}}{M_{25} \; \gamma_{15.6}} \quad \text{or} \quad \left(G_{60/60} = \frac{G_{77/77} \; \gamma_{w77}}{M_{77} \; \gamma_{w60}}\right) \tag{3}$$

where γ_{w25} (γ_{77}) = density of water at 25° C (77° F) = 0.9970 g/ml
$\gamma_{w15.6}$ (γ_{w60}) = density of water at 15.6° C (60° F) = 0.9988 g/ml
M_{25} (M_{77}) = multiplier from Table I-3, I-4, or I-5.

Assume the specific gravity of an asphalt cement at 77/77° F is 1.003. From Table 1-4, M for 77° F is 0.9941 (Column A). The specific gravity at 60/60° F is then

$$\frac{1.003 \; (0.9970)}{0.9941 \; (0.9988)} = 1.007$$

I.03 HORIZONTAL TANK MEASUREMENTS.—

Many asphalt containers for transporting and storage are cylindrical tanks in a horizontal position. Measurements of the volume of material in the tank consists of measuring its depth. Table I-6 gives quantities in terms of percent of the total capacity based on the depth in percent of the diameter.

Volumes of asphalt are usually expressed in litres (gal), and the formula for determining the capacity of a cylindrical tank is

$$\text{V} = 785 \; D^2 \; L \quad (V = 5.88 \; D^2 L) \tag{4}$$

where V = volumes, litres (gal)
D = diameter of interior of tank, m (ft.)
L = length of interior of tank, m (ft.)

Example: (U.S. Customary Units)

Assume a horizontal tank has a capacity of 12,740 gal and a diameter of 9 ft 6 in. If the depth of the asphalt in the tank is 7 ft 3.5 in. the percent depth filled is

$$\frac{7.29 \; (100)}{9.5} = 76.8 \text{ percent}$$

By interpolating in Table I-6, the volume of asphalt in the tank is

$$\frac{12,740 \; (82.40)}{100} = 10,498 \text{ gal}$$

When the tank is more than half full, as in the preceding example, it may be more desirable to determine the volume of the unfilled space and deduct this from the capacity. Using the above example, the computations would be:

$$12,740 - 12,740 \; \frac{(100-82.40)}{100} = 12,740 - 2,242 = 10,498 \text{ gal}$$

Table I-6. Percent Capacities for Various Depths of Cylindrical Tanks in Horizontal Position

Percent Depth Filled	Percent of Capacity	Percent Depth Filled	Percent of Capacity	Percent Depth Filled	Percent of Capacity	Percent Depth Filled	Percent of Capacity
1	0.20	26	20.73	51	51.27	76	81.50
2	0.50	27	21.86	52	52.55	77	82.60
3	0.90	28	23.00	53	53.81	78	83.68
4	1.34	29	24.07	54	55.08	79	84.74
5	1.87	30	25.31	55	56.34	80	85.77
6	2.45	31	26.48	56	57.60	81	86.77
7	3.07	32	27.66	57	58.86	82	87.76
8	3.74	33	28.84	58	60.11	83	88.73
9	4.45	34	30.03	59	61.36	84	89.68
10	5.20	35	31.19	60	62.61	85	90.60
11	5.98	36	32.44	61	63.86	86	91.50
12	6.80	37	33.66	62	65.10	87	92.36
13	7.64	38	34.90	63	66.34	88	93.20
14	8.50	39	36.14	64	67.56	89	94.02
15	9.40	40	37.39	65	68.81	90	94.80
16	10.32	41	38.64	66	69.97	91	95.55
17	11.27	42	39.89	67	71.16	92	96.26
18	12.24	43	41.14	68	72.34	93	96.93
19	13.23	44	42.40	69	73.52	94	97.55
20	14.23	45	43.66	70	74.69	95	98.13
21	15.26	46	44.92	71	75.93	96	98.66
22	16.32	47	46.19	72	77.00	97	99.10
23	17.40	48	47.45	73	78.14	98	99.50
24	18.50	49	48.73	74	79.27	99	99.80
25	19.61	50	50.00	75	80.39		

Table I-7. Weights and Volumes of Asphalt Materials (Metric Units) (Approximate)

Type and Grade	Kilograms per Litre	Kilograms per Barrel*	Cubic Metres per Tonne	Barrels* per Tonne
MC-30	0.935	149	1.068	6.7
RC-, MC-, SC-70	0.947	150	1.056	6.6
RC-, MC-, SC-250	0.959	153	1.039	6.5
RC-, MC-, SC-800	0.983	156	1.022	6.4
RC-, MC-, SC-3000	0.995	158	1.005	6.3
Emulsified Asphalts	0.995	158	1.005	6.3

*One barrel equals 159 litres.
NOTES: Since the specific gravity of asphaltic materials varies, even for the same type and grade, the weight and volume relationships shown above are approximate and should be used only for general estimating purposes. Where more precise data are required, they must be computed on the basis of laboratory tests on the specific product.
The approximate data shown above are for materials at 15.6° C.

Table I-8. Weights and Volumes of Asphalt Materials (Approximate)

Type and Grade	Pounds per Gallon	Pounds per Barrel*	Gallons per Ton	Barrels* per Ton
MC-30	7.8	328	256	6.1
RC-, MC-, SC-70	7.9	332	253	6.0
RC-, MC-, SC-250	8.0	337	249	5.9
RC-, MC-, SC-800	8.2	343	245	5.8
RC-, MC-, SC-3000	8.3	349	241	5.7
Emulsified Asphalts	8.3	349	241	5.7

*One barrel equals 42 U.S. gallons.
NOTES: Since the specific gravity of asphaltic materials varies, even for the same type and grade, the weight and volume relationships shown above are approximate and should be used only for general estimating purposes. Where more precise data are required, they must be computed on the basis of laboratory tests on the specific product.
The approximate data shown above are for materials at 60° F.

Table I-9. Weight Per Cubic Metre of Dry Mineral Aggregates of Different Specific Gravity and Various Void Contents

	Specific Gravity	Voids – Percent								
		15	20	25	30	35	40	45	50	55
KILOGRAMS PER CUBIC METRE	2.0	1700	1600	1500	1400	1300	1200	1100	1000	900
	2.1	1785	1680	1575	1470	1365	1260	1155	1050	945
	2.2	1870	1760	1650	1540	1430	1320	1210	1100	990
	2.3	1955	1840	1725	1610	1495	1380	1265	1150	1035
	2.4	2040	1920	1800	1680	1560	1440	1320	1200	1080
	2.5	2125	2000	1875	1750	1625	1500	1375	1250	1125
	2.6	2210	2080	1950	1820	1690	1560	1430	1300	1170
	2.7	2295	2160	2025	1890	1755	1620	1485	1350	1215
	2.8	2380	2240	2100	1960	1820	1680	1540	1400	1260
	2.9	2465	2320	2175	2030	1885	1740	1595	1450	1305
	3.0	2550	2400	2250	2100	1950	1800	1650	1500	1350
	3.1	2635	2480	2325	2170	2015	1860	1705	1550	1395
	3.2	2720	2560	2400	2240	2080	1920	1760	1600	1440

1. The Specific Gravity of commonly used road construction aggregates normally is within these ranges:

Granite	2.6-2.9	Sand (Quartzite)	2.5-2.7	Blast Furnace Slag	2.0-2.5
Gravel	2.5-2.7	Sandstone	2.0-2.7		
Limestone	2.1-2.8	Traprock	2.7-3.2		

2. Data contained in this table are applicable to dry mineral aggregates in either the loose or compacted state, and the void content should be selected accordingly. Preferably, both the void content and specific gravity should be determined in the laboratory.
3. The formula for computation of data in table above is:

$$W = 1000 \times \frac{G(100 - V)}{100}, \text{kg/m}^3$$

where W = Weight per cubic metre
G = Specific gravity
V = Air void content, percent

Table I-10. Weight Per Cubic Foot and Per Cubic Yard of Dry Mineral Aggregates of Different Specific Gravity and Various Void Contents

	Specific Gravity	VOIDS—PERCENT							
		15	20	25	30	35	40	45	50	55
POUNDS PER CUBIC FOOT	2.0	106.1	99.8	93.6	87.4	81.1	74.9	68.6	62.4	56.2
	2.1	111.4	104.8	98.3	91.7	85.2	78.6	72.1	65.5	59.0
	2.2	116.7	109.8	103.0	96.1	89.2	82.4	75.5	68.6	61.8
	2.3	122.0	114.8	107.6	100.5	93.3	86.1	78.9	71.8	64.6
	2.4	127.3	119.8	112.3	104.8	97.3	89.9	82.4	74.9	67.4
	2.5	132.6	124.8	117.0	109.2	101.4	93.6	85.8	78.0	70.2
	2.6	137.9	129.8	121.7	113.6	105.5	97.3	89.2	81.1	73.0
	2.7	143.2	134.8	126.4	117.9	109.5	101.1	92.7	84.2	75.8
	2.8	148.5	139.8	131.0	122.3	113.6	104.8	96.1	87.4	78.6
	2.9	153.8	144.8	135.7	126.7	117.6	108.6	99.5	90.5	81.4
	3.0	159.1	149.8	140.4	131.0	121.7	112.3	103.0	93.6	84.2
	3.1	164.4	154.8	145.1	135.4	125.7	116.1	106.4	96.7	87.0
	3.2	169.7	159.7	149.8	139.8	129.8	119.8	109.8	99.8	89.9
POUNDS PER CUBIC YARD	2.0	2860	2700	2530	2360	2190	2020	1850	1680	1520
	2.1	3010	2830	2650	2480	2300	2120	1950	1770	1590
	2.2	3150	2970	2780	2590	2410	2220	2040	1850	1670
	2.3	3290	3100	2910	2710	2520	2330	2130	1940	1740
	2.4	3440	3240	3030	2830	2630	2430	2220	2020	1820
	2.5	3580	3370	3160	2950	2740	2530	2320	2110	1900
	2.6	3720	3500	3290	3070	2850	2630	2410	2190	1970
	2.7	3870	3640	3410	3180	2960	2730	2500	2270	2050
	2.8	4010	3770	3540	3300	3070	2830	2590	2360	2120
	2.9	4150	3910	3660	3420	3180	2930	2690	2440	2200
	3.0	4300	4040	3790	3540	3290	3030	2780	2530	2270
	3.1	4440	4180	3920	3660	3400	3130	2870	2610	2350
	3.2	4580	4310	4040	3770	3500	3230	2970	2700	2430

1. The Specific Gravity of commonly used road construction aggregates normally is within these ranges:

Granite	2.6-2.9	Sand (Quartzite)	2.5-2.7	Blast Furnace Slag	2.0-2.5
Gravel	2.5-2.7	Sandstone	2.0-2.7		
Limestone	2.1-2.8	Traprock	2.7-3.2		

2. Data contained in this table are applicable to dry mineral aggregates in either the loose or compacted state, and the void content should be selected accordingly. Preferably, both the void content and specific gravity should be determined in the laboratory.
3. The formulas for computation of data in table above are:

$$W = 62.4 \times \frac{G(100-V)}{100} = 0.624\, G\, (100-V),\ \text{lb/ft}^3$$

and

$$W = 27 \times 62.4 \times \frac{G(100-V)}{100} = 16.85\, G(100-V),\ \text{lb/yd}^3$$

Where: W = Wt. per cu ft or cu yd
G = Specific gravity
V = Air void content, percent

Index

A

AASHTO and ASTM designated specifications for asphalt materials
  cutback and emulsified asphalts, 3
  PM-1 guideline, 37-38
  PM-2 guideline, 59
  RM-1 guideline, 48
  RM-2 guideline, 54
  RM-3 guideline, 63
Acceptance requirements in PM-1 guideline, 42-43
Aeration in cold mix construction
  basic information, 26
  RM-1 guideline, 50
  RM-2 guideline, 56
Aggregate
  choke, 33
  coarse, in Hveem method, 80
  combined, in Hveem method, 81
  fine, in Hveem method, 79
  gradations in maintenance mixes, 59, 63
  hard to coat, 8
  requirements for use, 6-8
  spreaders for, 14
  surface area calculation of, 76-77
  tests, 8
American Association of State Highway and Transportation Officials, methods of
  M82, 3, 37, 48, 54, 59, 63
  M140, 3, 37, 38, 48, 54, 59, 63
  M208, 3, 37, 38, 48, 54, 59, 63
  T2, 35, 46, 52, 58, 62
  T11, 36
  T19, 36, 46, 53, 58, 63
  T27, 36, 46, 53, 58, 63
  T37, 36
  T40, 35, 46, 52, 58, 62
  T90, 8, 53
  T96, 8, 36, 46, 58, 63
  T110, 26, 51
  T168, 35
  T176, 8, 36, 46, 53, 58, 63
  T245, 9
  T246, 9
  T247, 9
American Society for Testing and Materials, standards of
  C29, 36, 46, 53, 58, 63
  C117, 36, 147
  C127, 74, 147
  C128, 147
  C131, 8, 36, 46, 58, 63
  C136, 36, 46, 53, 58, 63, 147
  D75, 35, 46, 52, 58, 62
  D140, 35, 46, 52, 58, 62
  D242, 38
  D424, 8, 53
  D546, 36
  D977, 3, 37, 38, 48, 54, 59, 63
  D979, 35
  D1188, 87, 94, 113, 116, 152
  D1461, 26, 51, 139
  D1559, 9, 110, 111, 115, 146, 151
  D1560, 9
  D1561, 9, 86
  D2026, 3, 48, 59, 63
  D2027, 3, 37, 48, 54, 59, 63
  D2041, 151, 158
  D2216, 84, 108
  D2397, 3, 37, 38, 48, 54, 59, 63
  D2419, 8, 36, 46, 53, 58, 63, 123
  D2726, 87, 94, 113, 116, 152
  D2950, 44
  D3515, 123
  D4311, 37, 47, 165
  E11, 159

178 Index

Anionic emulsified asphalt, definition, grades, and referenced specifications, 3
Applications of asphalt cold mixes, limitations of, 1-2
Arbor press, 111, 146
Asphalt
    basic information, 3-6
    characteristics of residue of, 5
    consistency, 5
    cutback, 3
    dispersion, 83
    emulsified, 3
    guide to uses, 4
    rate of curing, 5-6
    selecting the proper asphalt, 3
    temperature-volume correction tables, 166-170
    types of asphalt, 3
    typical temperatures of asphalt for cold mixed construction, 7
    weight and volume relationships, tables, 173
Asphalt application for mixed-in-place construction, determining rates of, 20-24
Asphalt binder
    PM-1, 37-38
    PM-2, 59
    RM-1, 48
    RM-2, 54
    RM-3, 63
Asphalt cold mix, introduction to, 1-2
    advantages, 1
    limitations, 1-2
    roadbed preparation, 2
Asphalt distributors, 14
Asphalt cutbacks and emulsified asphalts, typical temperatures for mixing in pugmill and spraying, and safety cautions, 162-164
Asphalt mixture in hauling trucks, random sampling plans for, 66
Asphalt primer grades in RM-1 guideline, 45
Asphalt seal coat in RM-2 guideline, 56

B

Basic information for asphalt cold mix, 1-10
    aggregates, 6-8
    asphalt, 3-6
    introduction, 1-2
    mix design, 9-10
Blade grading, transverse tolerance, 25
Blade-mix construction details, 25
Blade mixing
    operation and asphalt application, 25
    in RM-1, 49
    in RM-2, 55

C

Cationic emulsified asphalt, definition, grades and referenced specifications, 3
Central plant-mix, 29-33
    construction, 32-33
    general, 29
    mixing equipment, 29-31
    spreading equipment, 31
Choke aggregate use, 33
C.K.E., oil ratio, 78, 82, 126
C.K.E. test, 75-82
    aggregate surface constants, 119-122
    equipment for, 75-76
    procedure, 77
Coating test, 107-110
Cohesiometer
    calibration of, 101, 139
    illustrated, 104, 136
    test, 101, 139
    value, 103, 142
Cold asphalt plant-mix guideline PM-1, 35-44

Compacting equipment, 16-19
  pneumatic-tired rollers, 16
  steel-wheeled rollers, 17
  vibratory rollers, 19
Compaction
  fluids content, and, 74, 86
  optimum water content at, 110
  specimens, of, 87, 113, 130-132, 150
Construction specifications of PM-1 guideline, 41-44
  acceptance requirements, 42-43
  compacting the mixture, 42
  notes to the engineer, 43-44
  placing the mixture, 42
  preparing area to be paved, 41
  preparing the mixture, 41
  tack coat, 41
Construction specifications of RM-1 guideline, 48-51
  aeration of mixture, 50
  application of asphalt, 49
  blade-mixing, 49
  notes to the engineer, 51
  preparation of mineral aggregate for mixing, 48-49
  spreading and compaction, 50
  travel-mixing, 49-50
Continuous mix plant, 29, 30-31
Curing rates of cutbacks and emulsified asphalts, basic information, 5-6
Cutback asphalts
  basic information, 3
  definition, grades, types and referenced specifications, 3
  Hveem method, in, 123-143
  maintenance mixes, in, 58-65
  Marshall method, in, 144-162
  typical temperatures in pugmill mixing and spraying, 162-163
  safety precautions, 164

D

Density requirements for acceptance in PM-1 guideline, 42

Design criteria for asphalt-aggregate mixtures, 39, 105, 121, 143, 156
Determining asphalt application rates for mixed-in-place construction, 42
  control of asphalt, 24
  example, 21-24
  formulas, 21, 22, 23
Distributors, 14

E

Economical advantages of asphalt cold mix, 1
Empirical formulas for cold mix design, 9-10
Emulsified asphalt
  basic information, 3-6
  definitions, grades, types and referenced specifications, 3
  Hveem method, in, 71-105
  maintenance mixes, in, 58-65
  Marshall method, in, 106-122
  optimum content determination, 104, 120
  typical temperatures in pugmill mixing and spraying, 162-163
Equipment
  central plant mixing, for, 29-31
  compacting, 16-19
  laboratory, 93, 95, 96, 97, 104, 107-108, 110-111, 115, 128, 132-133, 145-146, 151
  mixed-in-place mixtures, for, 11-19
  mixing, 11-14
  spreading, 14-16

F

Fluids
  compaction, optimum for, 86
  content of mix, 83-86
Formulas
  cohesiometer value, 103, 142
  cylindrical tank capacity, 171
  determining asphalt application rate, 21, 22, 23

180 Index

dry weight of aggregate
  per cubic foot, 175
  per cubic metre, 174
  per cubic yard, 175
  specimens, 114, 129, 149
empirical, mix design, 9, 10
M_r, 93
percent moisture pick-up, vacuum saturation, 96
resistance R-Value, 98
resistance R_t-Value, 98
specific gravity, 165, 171
stabilometer S-Value, 101, 142
temperature-volume, asphalt, 165
trial emulsified asphalt content, 83, 107
windrows
  asphalt application rate, 22
  forward speed of distributor, 23
  volume, 21
  weight of materials, 21
Forward speed of distributor or mixer needed to ensure asphalt application rate, 23

G

General requirements of cold plant-mix guideline PM-1, 35-37
  basis of payment, 37
  equipment, 35
  methods of measurement for pay items, 37
  methods of testing, 36
  placement limitations, 36
  safety, 36
  sampling, 35
  traffic control, 36
General requirements of mixed-in-place courses guideline RM-1, 45-47
  basis of payment, 47
  equipment, 46
  method of measurement for pay items, 47
  methods of testing, 46
  preparation of road surface, 45
  safety, 47
  sampling, 46
  traffic control, 47
  weather, 47
General requirements of road-mixed asphalt courses for base and surface (sand or soil) guideline RM-2, 52-54
  basis of payment, 54
  equipment, 52
  method of measurement for pay items, 54
  method of testing, 54
  preparation of roadway, 52
  safety, 53
  sampling, 53
  traffic control, 53
  weather, 53
Guideline PM-1 for cold asphalt plant mix, 35-44
  construction, 41-43
  general requirements, 35-37
  materials, 37-40
  notes to the engineer, 43-44
Guideline PM-2 for plant-mixed asphalt stockpile maintenance mixtures, 58-61
  basis of payment, 60
  equipment, 58
  method of measurement, 60
  methods of testing, 58, 59
  mineral aggregate gradations, 59
  notes to the engineer, 61
  preparation of mixture, 60
  sampling, 58
  stockpiling, 60
Guideline RM-1, mixed-in-place courses, 45-51
  construction, 48-50
  general requirements, 45-47
  materials, 48
  notes to the engineer, 51
Guideline RM-2, road-mixed asphalt courses for base and surface (sand or soil), 52-57
  construction, 55-56

general requirements, 52-54
materials, 54
notes to the engineer, 57
Guideline RM-3 for mixed-in-place asphalt stockpile maintenance, 62-65
  basis of payment, 65
  equipment, 62
  method of measurement, 65
  methods of testing, 62-63
  mineral aggregate gradations, 63
  notes to the engineer, 65
  preparation of mixture for blade mixing, 64
  preparation of mixture for travel mixing, 64
  sampling, 62
  stockpiling, 64

H

Hard-to-coat aggregate, improvement measures, 8
Haul trucks, asphalt, random sampling plans for, 66
Hveem cohesiometer, illustrated, 104, 136
Hveem mix design, modified
  apparatus for C.K.E. test in, 84
  coarse material in, 80
  combined aggregate in, 81
  cutback asphalt, for, 123-143
  emulsified asphalt, for, 71-105
  fine material in, 79
  oil ratio chart in, 82
Hveem stabilometer, 97, 135
Hveem test procedures for cold mix design
  cutback asphalts, 123-143
  emulsified asphalts, 71-105

I

Inspecting and sampling for quality control of mixed-in-place construction, 27

J

Job-mix formulas for base and surface mixtures, PM-1 guideline, 38-40
  limits, 39

L

Linear measurement covered by tank of any capacity for various widths and rates of application, 171-172
Los Angeles abrasion test, 8
Low or non-polluting advantages of asphalt cold mixes, 1

M

Maintenance (stockpile) guidelines
  PM-2, 58-61
  RM-3, 62-65
Marshall mix-design, modified
  cutback asphalt, for, 144-161
  emulsified asphalt, for, 106-122
Materials for asphalt cold mixes
  aggregates, 6-8
  asphalt, 3-6
Material specifications of PM-1 guideline, 37-40
  asphalt-aggregate base and surface courses
    job-mix formula limits, 38-39
    test criteria, 40
  mineral aggregate materials for base and surface courses, 38
  notes to the engineer, 43-44
Material requirements in RM-1 guideline, 48
  asphalt binder, 48
  mineral aggregate, 48
Material requirements in RM-2 guideline, 54
  asphalt binder, 54
  mineral aggregate, 54
Methods of testing in guidelines
  PM-1, 36, 37-38
  PM-2, 58, 59
  RM-1, 46, 48

RM-2, 53, 54
RM-3, 62-63
Mineral aggregate test methods in guideline PM-1, 36
Mix-design methods for cutback asphalts
   empirical formula, 10
   general, 9
   Hveem test procedures, 123-143
   Marshall test procedures, 144-161
Mix-design method for emulsified asphalts
   empirical formula, 9
   general, 9
   Hveem test procedures, 71-105
   Marshall test procedures, 106-122
Mixed-in-place construction, 20-27
   aeration, 25
   control of asphalt, 24
   determining asphalt application rates for, 20-24
   inspecting and sampling, 27
   mixing, 24-25
   spreading and compacting, 26-27
   windrows, 20
Mixed-in-place equipment, 11-19
   compacting, 16-19
   mixing, 11-14
   spreading, 14-16
Mixed-in-place guidelines
   RM-1, 42-51
   RM-2, 52-57
   RM-3, 62-65
Mixer forward speed to determine asphalt application rate, formulas for, 21, 22, 23
Mixing in cold mix construction by
   blade mixing, 25
   rotary mixing, 24-25
   travel-plant mixing, 25
Moisture vapor susceptibility test, 138
   criteria for cutback asphalt mixes, 143
   illustrated, 141
Motor graders, 13, 14-16

mixing, 13
spreading, 14, 16
M_r calculation, 93
M_r measurements, 90-94
M_r pressure regulator, 90, 93
M_r recording meter, 90, 93
M_r test, 89-95
   measurements in procedure, 90-94
   yoke in, 90, 91, 92

N

Notes to the engineer in guidelines
   PM-1, 43-44
   PM-2, 61
   RM-1, 51
   RM-2, 65
   RM-3, 57

O

Oil ratio chart, Hveem mix design, 82
Oil ratio, C.K.E., 78, 126

P

Patching mix guidelines
   PM-2, 58-61
   RM-3, 62-65
Pavement cores for density and thickness with referenced specification in PM-1 guidelines, 42
Pavement sampling by random sampling plan, 68-69
Pavers, central mix plant, 29
Pay items, basis and method of measurements in guidelines
   PM-1, 37
   PM-2, 60
   RM-1, 47
   RM-2, 54
   RM-3, 65
Placement limitation in PM-1 guideline, 36
Plant-mix equipment, 29-31
   spreading, 31, 32-33
   compacting, 32-33

Plasticity index criteria, 8
PM-1 cold plant-mix guideline, 35-44
PM-2 plant-mixed asphalt stockpile maintenance mixtures, guideline, 58-61
Pneumatic-tired rollers, 16-17
Preparation of mineral aggregate in
 RM-1 guideline, 48-49
 RM-2 guideline, 55
Preparation of mixture in central plants, 32
Preparation of road surface in RM-1 guideline, 45
Preparation of roadway in RM-2 guideline, 52
Preparing area to be paved in PM-1 guideline, 41
Pugmill mixture temperatures for cutbacks and emulsified asphalts, 162-163
 Caution for safe handling, 164

Q

Quality control of asphalt cold mixes, 2

R

Random sampling plans, 66-70
 asphalt mixtures in hauling trucks, 66-67
 compacted pavement sampling, 68
 random numbers for general procedure, 69-70
 schematic diagram illustrating lot, sample, subsample, and sample unit, 67
Resilient Modulus
 See M_r
Resilient Modulus apparatus, 91
Resistance R-Value test, 96
RM-1 mixed-in-place courses guideline, 45-51
RM-2 road-mixed asphalt base and surface courses (sand or soil), guideline, 52-57

RM-3 mixed-in-place asphalt stockpile maintenance mixtures, guideline, 62-65
Roadbed preparation for asphalt cold mixes, 2, 20
Rollers
 pneumatic-tired, 16-17
 steel-wheeled tandem and three-wheel, 17-19
 vibratory, 19
Rotary-mixing, construction details, 11-12
Rotary-type mixers, 11
R-value, 74
 chart for correcting, 100
 chart for determining, 99
 formula, 98

S

Safety
 cautions in handling asphalt cutbacks, 164
 precautions in PM-1, 36
 precautions in RM-1, 47
 precautions in RM-2, 53
Sampling asphalt in trucks, 66-67
Sampling compacted pavement, 68-69
Sampling materials in guidelines
 PM-1, 35
 PM-2, 58
 RM-1, 46
 RM-2, 52
 RM-3, 62
Sand-asphalt road-mix, guideline RM-2, 52-57
Sand equivalent test, use of, 8
Sand equivalent test value, 38, 54, 59, 63
Seal coat in RM-2 guideline, application of, 56
Selecting the proper asphalt, 3-6
 characteristics of residue, 5
 consistency, 5
 guide to uses of asphalt, 4

rate of curing, 6-7
typical asphalt temperatures for construction, 5
Smoothness requirements for surface, PM-1 guideline, 43
Soil-asphalt road-mix guideline RM-2, 52-57
Specifications referenced for cutback and emulsified asphalts, 3
Spraying temperatures for cutback and emulsified asphalts and safety caution, 162-164
Spreader types for central plant-mix, 16
Spreading equipment, 14-16, 31
   discussion, 16, 31
   motor grader requirements, 14-16
Spreading and compacting mixed-in-place construction, 26-27
   RM-1, 50
   RM-2, 56
Spreading and compacting plant mix, considering lift thicknesses, 33
Stability, Marshall
   correlation ratios, 119
   criteria for cutback mixes, 156
   criteria for emulsion mixes, 121
   percent loss of, 116, 120
   percent retained, 156
   soaked, test, 117
   testing for, 117, 152
Stabilometer
   displacement of, 96-98, 138
   illustrated, 97, 135
   value
      calculating, 101, 142
      chart for correcting, 102
      criteria for cutback asphalt mixes, 143
      criteria for emulsified asphalt mixtures, 105
Stationary asphalt plants, 29-31
Steel-wheeled rollers, 17-19
Stockpile (maintenance) guidelines
   PM-2, 58-61

   RM-3, 62-65
Stockpiling of maintenance mixtures, 60, 64
Surface area of aggregate, 76, 124
Surface capacity test, 77
Surface constant, Hveem method
   coarse material, for, 80
   combined material, for, 81
   fine material, for, 79
Surface moisture limitations of asphalt cold mixes, 1-2
S-Value testing, 98-101
   formula, 101, 142
Swell test, 133
   apparatus, 134
   criteria for cutback asphalt mixes, 143

T

Tack coat, 41
Temperature-volume corrections
   asphalt materials, 165-169
   emulsified asphalts, 165, 169
Test criteria for asphalt-aggregate mixtures, 39, 105, 121, 143, 156
Tire pickup prevention, 27, 33
Traffic control in guidelines
   PM-1, 36
   RM-1, 47
   RM-2, 53
Travel plants
   construction details, 25
   general, 13-14
Travel-mixing operation and asphalt application guidelines
   RM-1, 49-50
   RM-2, 56
   RM-3, 64
Trial mixes, emulsified asphalt content in Hveem, 83
Trucks hauling asphalt, random sampling plan for, 66
Typical asphalt temperatures for cold mix construction, 7

Typical temperatures for uses of cutbacks and emulsified asphalts, tables, 162-163

V

Vacuum saturation test, 95-96
Versatility of asphalt cold mixes, 1
Vibratory rollers, 19
Viscosity, consistency, basic information, 5
Volume of windrows, formula for, 21

W

Water distributor, 14
Weather
  asphalt cold mix limitations, 1
  PM-1 guideline, 36
  RM-1 guideline, 47
  RM-2 guideline, 53
Weight and volume relations of asphalt materials, 173
Weight of
  asphalt materials, 173
  dry mineral aggregates
    differing gravities and voids per cubic metre and cubic foot, 174-175
Windrows, 20
  blade mixing and, 20, 23
  flattened, 23
  formula for volume determination, 21
  moving of, 25

LaVergne, TN USA
17 February 2011
216941LV00003B/1/P